Telecommuting

The Future Technology of Work

Telecommuting

The Future Technology of Work

Thomas B. Cross

Marjorie Raizman

DOW JONES-IRWIN

Homewood, Illinois 60430

CHICAGO RESEARCH AND TRADING GROUP, LTD.
440 SOUTH LA SALLE STREET
CHICAGO, ILLINOIS 60605

We recognize that certain terms in this book are
trademarks, and we have made every effort to print
these throughout the text with the capitalization
and punctuation used by the holder of the trademark.

ISBN 0-87094-645-5

Library of Congress Catalog Card No. 86–70514

Printed in the United States of America

1 2 3 4 5 6 7 8 9 0 MR 3 2 1 0 9 8 7 6

To Mr. R. Craig Blackman of Barry University, and Mr. Peter K. Dallow of the city of Fort Collins, Colorado who gave us the opportunity to excell.

Foreword

Telecommuting means working whenever and from wherever the work is needed. Telecommuting opens the work horizon by creating an office without walls, an office that, in fact, can extend to an airplane passenger seat, satellite work center, or home dining room located anywhere in the world. Telecommuting can also take place in "flextime," giving both workers and employers an opportunity to operate during hours that are best for them.

Although telecommuting depends on current technology, it is basically a social phenomenon that many people consider to be an updated version of the age-old "cottage industry" concept. As with most technological innovation, there has been considerable lag time between the introduction of telecommuting technology and its broad acceptance in business, government, and elsewhere. Those innovations that will have the greatest impact on the remote work picture are just beginning to evolve. Robotics, smart appliances, and advanced electronic communications systems are among the promising new possibilities.

ACKNOWLEDGMENTS

We would like to thank Mr. Alan W. Harris, Mr. Andrew McKay, Mr. Ronald Kauffman, and Mr. Marc Raizman for their friendship, inspiration, and guidance. We would also like to thank Ms. Lynn Parrill for her editorial assistance.

Contents

Telecommuter Idea Exchanges. Telecommuting as a Decision Support System. Expert Systems. Trends in Telecommuting.

Telecommuting Conferences. Telecommuting Support Organizations. Telecommuting Newsletters. Telecommuting Books and Related Issues. Smart Home Products. On-Line Database Services. Audio Bridging Companies: *Bridge Locations by Service.* Telecommuting Business Case. Major Products/Services Suggested for Intelligent Buildings: *Intelligent Building Management. Intelligent Command and Control. Information Technologies. Intelligent Resources.* The Grapefruit Diet: *Implementing Electronic Communications.*

I

Overview, Driving Factors, Implementation, and Human Factors

1

Telecommuting Overview

An estimated 13 million people currently work at home at least part time.[1] As that number grows and computers become a more important part of the home working scene, telecommuting in its many forms is emerging as a significant social and economic trend. In fact, futurists forecast that 15 million people will be telecommuting two or three days each week by 1990.[2] Thus the electronic cottage, foreseen as the evolving home setting for office work, is becoming one of the more talked-about business office scenarios.

Prospects for the growth of telecommuting on all business levels could be even rosier, however, given the current pace of technological developments in the computer and telecommunications industries and the fact that 60 percent of American jobs currently involve information handling.[3] Electronic Services Unlimited (ESU) of New York estimated in 1984 that some 7.2 million workers were potential telecommuters, although only about 100,000 people were engaged in formal and informal remote-work programs for 450 companies.[4]

As with all such trends, employers and employees are asking how telecommuting can make their work more effective and efficient, how they can best profit from it, and what its pitfalls are. Chapter 1 provides an overview of telecommuting that answers the following questions: What is telecommuting? How does it benefit business and employees? Which tasks are suitable for remote work? Who telecommutes? How do telecommuters evaluate their experiences? What barriers are impeding the growth of remote work?

WHAT IS TELECOMMUTING?

Telecommuting means performing job-related work at a site away from the office, then electronically transferring the results to the office or to an-

3

FIGURE 1-1

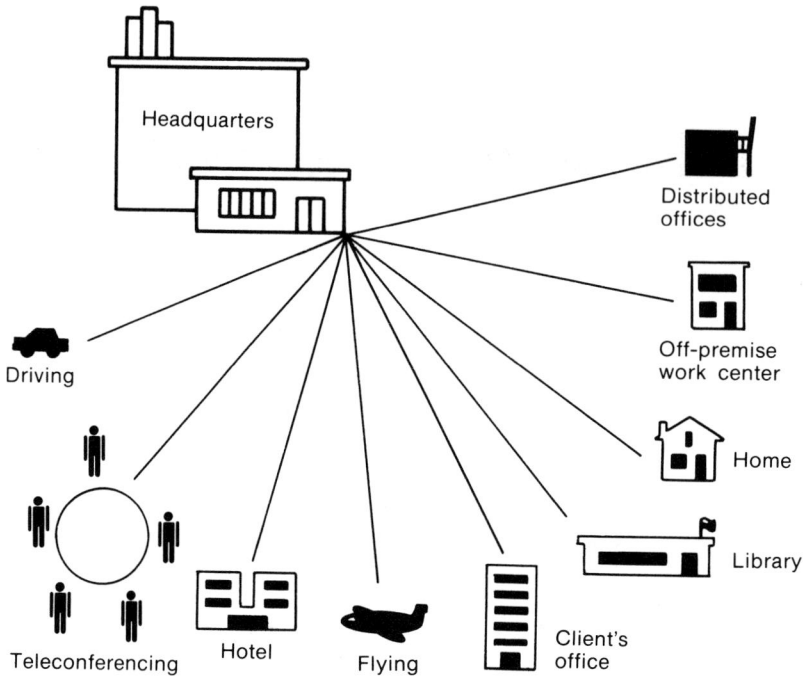

Telecommuting means working at home or elsewhere and transferring the work product electronically to the office.

other location. Telecommuting usually supplements other office activities, although many jobs can be handled entirely in this way. See Figure 1–1.

Among the popular terms that cover telecommuting, one hears: remote work, home office work, telework, location-independent tasks, and "home-distributed data processing."[5] Control Data Corporation, Minneapolis, calls its program for the handicapped "Homework,"[6] and federal agencies refer to telecommuting as "industrial homework."[7]

Remote work is often performed on a "dumb" terminal that is connected by modem and telephone line to a mainframe computer that processes the work. "Dumb" in this context means the terminal has limited calculating, processing, and storage capabilities. Remote work is also often performed and processed on a stand-alone microcomputer (also known as a personal computer or PC). In the latter instance, the completed task is transmitted over telephone lines (uploaded) to the company's computer facilities, or the disks on which the work is entered are

FIGURE 1-2

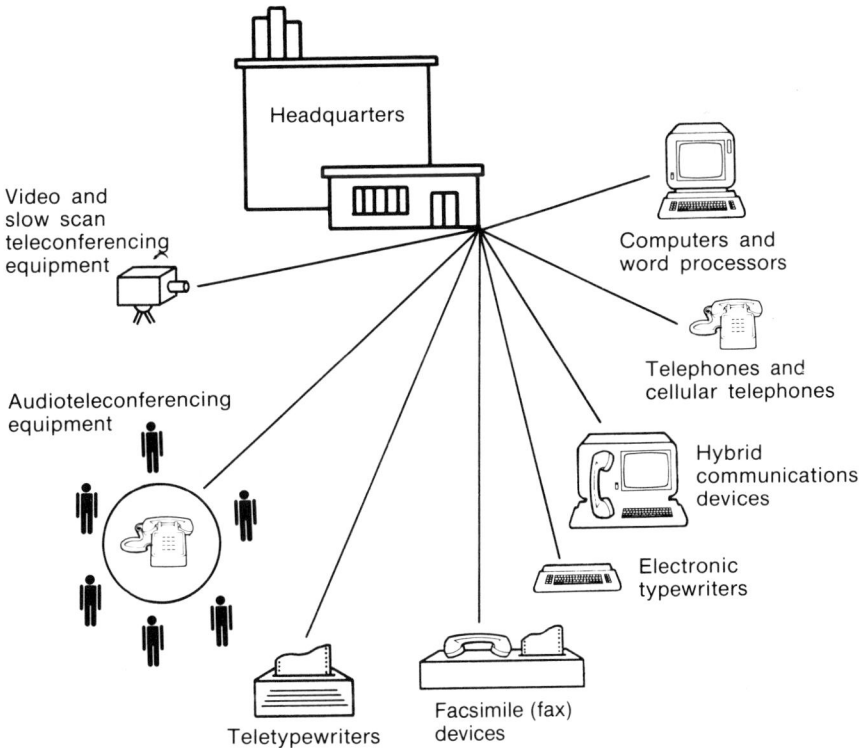

Video and
slow scan
teleconferencing
equipment

Audioteleconferencing
equipment

Teletypewriters

Facsimile (fax)
devices

Headquarters

Computers and
word processors

Telephones and
cellular telephones

Hybrid
communications
devices

Electronic
typewriters

Devices that can be used to telecommute.

carried or mailed to the office. The work is then transferred to the company's main computer. See Figure 1–2.

When remote terminals are connected to an organization's communications network, employees can use electronic mail or computer teleconferencing software systems to call up electronic files or database information onto their screens and work on it. The results can be transmitted back to the office, to a supervisor, or deposited in the appropriate file. Any number of people can be working remotely for an organization at the same time. Electronic mail and teleconferencing software systems are highly strategic telecommuting tools that are explained at length in later chapters. See Figure 1–3.

Telecommuting can be handled in other ways as well. Reporters have telecommuted for years by sending in their stories to newspaper city rooms via telephone, teletypewriter, or facsimile equipment. Sales representatives and customer service employees, when calling on cli-

FIGURE 1-3

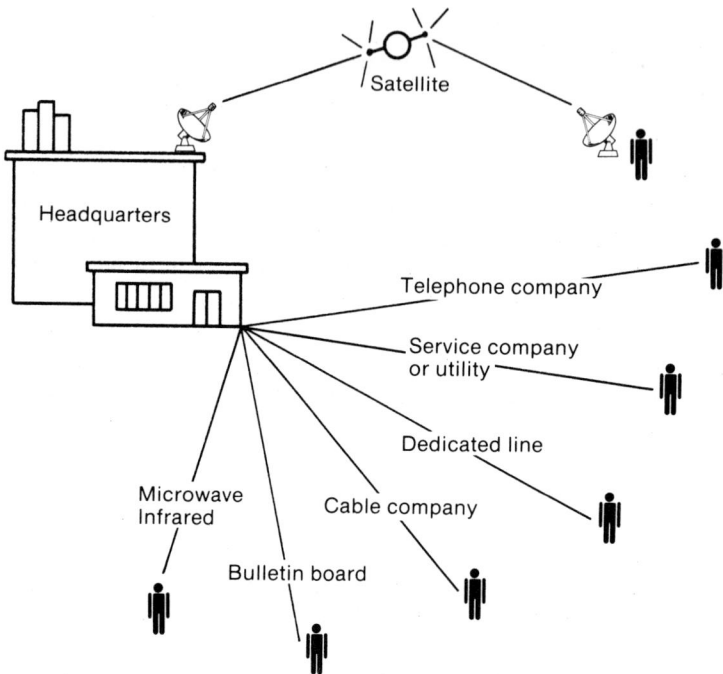

The telecommuter can connect to the office in a number of ways.

ents, regularly connect their portable computers to the client's telephones. This has enabled computer users to reach their companies' computers and query them on inventory information, product availability, current prices, and delivery schedules. Salespeople can then use the same linkage to place product orders and perform other related functions. See Figure 1–4.

Simple telephone conference calls can be used to transmit information (work) and coordinate activities, as well as to avoid costly face-to-face meetings. Electronic mail and computer teleconferencing systems also allow participants to send their work directly to other people or place it in electronic files which others can access. Today, not even telephone lines are required for sending and receiving information. Cellular radio, paging systems, and new over-the-air technologies make it possible for the employee to work from virtually any location.

Thus, for all practical purposes, remote work can be accomplished at the time and place one chooses, then transmitted from any appropriately equipped site. Depending on the nature of the material and the

FIGURE 1-4

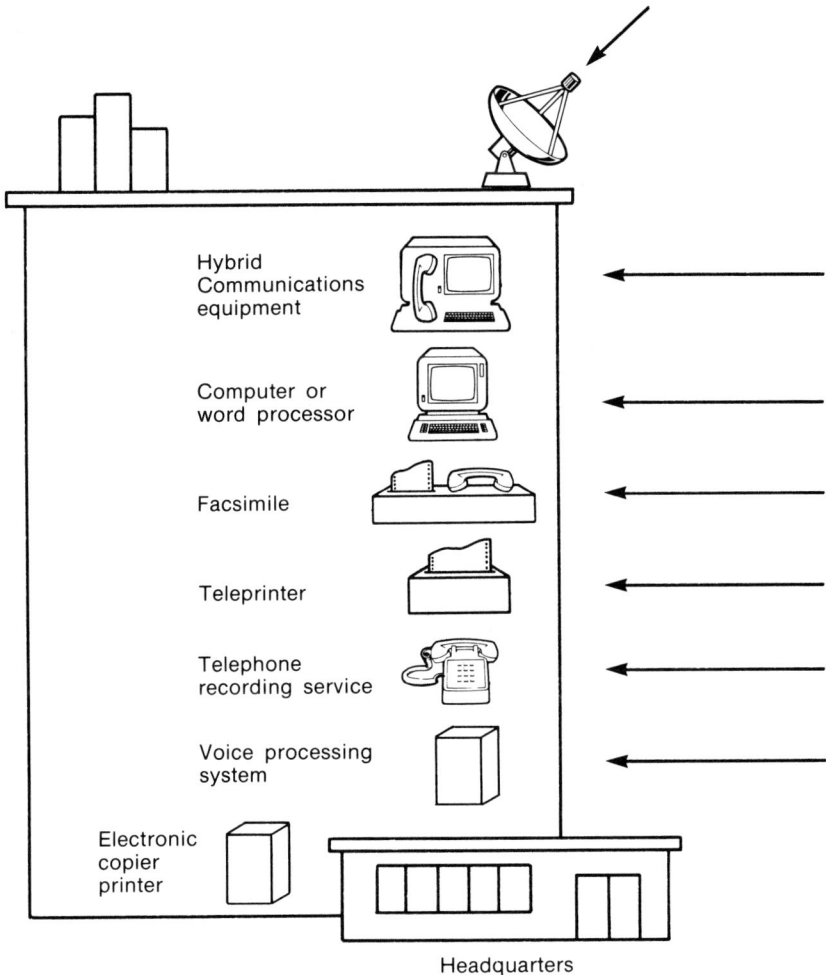

Hybrid
Communications
equipment

Computer or
word processor

Facsimile

Teleprinter

Telephone
recording service

Voice processing
system

Electronic
copier
printer

Headquarters

The office can receive the telecommuter's work on a number of devices.

transmission it requires, the site could be a telephone booth, a teleconferencing meeting, a client's office, in or out of town, around the country or the world. This way of operating is often called operating asynchronously, or being time and site independent.

Telecommuting does not necessarily mean working only at home or using a computer or telecommunications technology to link the office and home. Offices located at any distance from headquarters, at a neighborhood work center, hotel room, airplane, or telephone booth can be telecommuting sites. In addition, there are a number of ways to transfer the work product. These will be explained in the following chapters.

It should be noted that not all office tasks lend themselves to telecommuting. Inappropriate jobs are those that require employees to frequently interact face-to-face with associates or clients, to handle products, or to manage resources (microfiche or paper) that are not electronically based.

The principle behind telecommuting is the old "cottage industry" concept where workers do piecework at home then turn it in to the employer. But telecommuting today is influenced by many other factors. Primary among them is decentralization of the work process and the automated office, and management's need to process ever-increasing amounts of information. Currently, decentralization of office work has put most office tasks in the machine, and the machine—now located almost anywhere—enables employees to access databases, graphics, electronic mail, word processing, and other software system applications. The Resource Guide at the end of the book offers information on organizations, newsletters, magazines, and services that support telecommuters and their work.

WHAT ARE THE BENEFITS OF TELECOMMUTING?

Telecommuting can benefit every segment of the economy—directly and indirectly—by permitting work to be done where it is convenient and information to flow to and from choice locations during the most appropriate hours. These time, place, and networking options lead to increased production and reduce those costs related to tasks performed by an organization. Businesses, banks and governmental organizations, networked communities, educational systems, entrepreneurs, and individual employees are among the many that can benefit from this technology.

Telecommuting programs provide an additional bonus. They can improve the quality and increase the speed of an organization's communications and decision-making processes, enhancing its general overall productivity and reducing communications costs.

An organization can save money on location costs when some tasks are carried out in less costly facilities or in employees' homes. For example, the typical metropolitan area firm spends from $4,000 to $6,000 per worker for office space,[8] a sum that is two to three times more than the cost of supporting a telecommuter at home.[9] The home worker does not charge the employer for space used (although this may change), for heating, air conditioning, or for janitorial service. Thus, telecommuting can be used as a hedge against eventual shifts in urban office and land prices.[10]

Although productivity among telecommuters often initially declines, it increases from 2 percent to 40 percent after the period of adjustment is over, providing an organization with 20 to 300 percent worker-productivity increases, according to Jack Nilles of the University of Southern California.[11] A mid-1984 survey indicates that telecommuters are 20 percent more productive than their office counterparts.[12] Such gains are significant for organizations that seek to cap the estimated 15 percent yearly cost increase of running an office.

Telecommuters are believed to produce more because they work during the hours they feel best and are most proficient, they work without interruption, and they put in longer hours.[13] Managers must, however, choose tasks that lend themselves to remote work and appropriately handle the technical and human-resource aspects of dealing with the remote worker.

In order to save on wages through telecommuting programs, some companies have extended their remote operations beyond national boundaries. Satellite Data Corporation (SDC), for example, has telecommuters working in the Caribbean for a fraction of New York City wages and is considering expanding its program to India. A well-educated but underemployed work force there might handle information at night for next-day delivery to SDC customers.[14]

Neodata, a large Boulder, Colorado, company handles thousands of magazine subscriptions and renewals daily. After removing money from incoming envelopes, subscription forms are air shipped to Ireland where workers enter subscriber information into terminals that are linked to the company's mainframe computer in Boulder.

When markets fluctuate and a company is uncertain of its long-range requirements, it can often better control building space, support facilities, and costly equipment by having staff members work remotely. This is particularly true during periods of business expansion. Or, when clientele are spread across the country, the organization may find it more cost effective to distribute customer support facilities among telecommuters than to develop a central office location that requires storage space, a cafeteria, and other facilities.[15]

Telecommuting helps organizations retain valued employees when operations are expanded and space constraints develop or when a change in location is required. In one instance, a midwestern company reported a saving of $130,000 in staff turnover when it integrated a telecommuting program to retain the services of 12 professionals. Current figures on the cost of employee relocation range from $30,000 to $50,000 in direct expenses for a home-owning worker, with surveys predicting that such costs might eventually climb to $90,000.[16] In another instance, a Connecticut-based insurance company that turned to telecommuting to retain key senior analysts and programmers expanded its project to include 25 professionals.[17] With continued success, it expects to increase that number to 200.

Organizations that rely on telecommuters can save money by sharing satellite office facilities that are located in unused schools, shopping centers, day-care, and other vacant facilities, etc. Likewise, dispersed or neighborhood work centers enable businesses to recruit specialized personnel who prefer to work near home and enjoy doing so part time or during unusual hours. Such personnel can be found among computer users' groups, computerized bulletin board system listings for specialized groups, or electronic information utilities such as H&R Block's CompuServe™ or the Reader's Digest's The Source™, as well as through newspaper advertisements.

Besides multiplying the organization's time and place options for handling tasks or communications, telecommuting enables organizations to obtain and control information by:

- Placing work within an electronic storage system so that it can be retrieved when supervisors or workers need it.
- Moving work to people, often saving them time that would be spent in face-to-face meetings.
- Enabling management to treat work objectives as part of a system rather than as the function of a location.
- Permitting businesses to be open additional hours.

Among the other benefits of telecommuting, organizations find there is:

- More work of a higher quality (lower unit labor costs) completed during shorter periods in flextime programs.[18]
- More diversified work accomplished (depending on contributors).
- Improvement in supervision. The manager's attention is focused on work rather than the employee's congeniality or appearance.[19]
- Opportunity to use underutilized but knowledgeable workers.
- Time saved in communicating. Using electronic mail in place of telephone calls is known to free about three hours for other office work each week.
- Improved morale among telecommuters and reduced occasional absences.
- Decreased office traffic.
- Fewer stress-related problems among managers. They can replace some face-to-face (time-consuming and/or confrontational) meetings with telecommuting.
- Possible elimination of costly benefits such as cafeteria service in home offices.

- More notice taken by the public of positions that are advertised as offering a telecommuting option.

Both employees and managers can benefit from remote-work situations when short-term employment and/or flexible working hours are desirable. While many workers have no alternative but to work unconventional hours, numerous organizations have costly mainframe computers that can be used after business hours when they are more efficient and less costly to operate.

In addition, telecommuting can serve both managers and employees because it:

- Allows specialists, the handicapped and homebound, or those too remote or busy to travel to participate in the work or projects.
- Creates greater mutual respect and trust when employer and employee establish clearly understood remote-work conditions and objectives. It also increases workers' loyalty to the company that places trust in them.
- Eliminates time lost in unnecessary social interaction and in meetings.
- Callers (clients, students, etc.) do not have to know the employee's location when call-forwarding and conference calling communications systems are used.
- Appearance, dress, and style of working go unobserved by supervisor and telecommuter.

Telecommuting provides workers, academics, and others with a number of practical, financial, psychological, and environmental benefits. By way of practical benefits, people generally report being pleased with having control over their time and place of work and being able to operate in a quiet setting that allows full concentration. This allows them increased productivity, more time both for themselves and their families, and the chance to make more money and increase their career opportunities.[20] In addition, home quietude can provide an escape from the currently popular, open-office design. Such offices are known for their lack of privacy, noise, "cubicle clutter," interruptions, and monitoring.

Among other practical benefits of remote-work programs, employees report the ability to access vast amounts of database information that applies to their work or interests, while it eliminates the travel costs and stress of fighting traffic, time lost in traveling, certain child-care frustrations, and maintaining a very well-dressed appearance. Telecommuting (being site independent) also provides employees with more flexibility in choosing living quarters and in being able to remain in a desirable job when they must relocate or the company must move.

For the elderly retired, remote work can provide an avenue to earn an income and keep mentally alert, one remote worker claims.[21] Where the system allows for initiative and innovativeness, the remote worker has increased chances for promotion.[22]

Telecommuting provides workers with a number of psychological benefits as well. Many people enjoy a sense of autonomy, perhaps the most important aspect of telecommuting, and consider it to be "a privilege, (and) an exercise in personal preference and commitment."[23] Workers also take satisfaction in helping to structure remote tasks, where given the chance, and achieve a balance between productivity and bottom-line responsibility. Another psychological benefit of working remotely is perceiving a greater importance in one's position within the system. The exception is found among people in the military and female professionals who are concerned with office visibility, according to a Cornell University survey.[24]

Some telecommuters find unusual advantages in their telecommuting arrangements. For example, Joyce and Pat Kelly, local government professionals in Fort Collins, Colorado, have installed his and hers computers at home so that they can work overtime hours there, together, at the jobs they enjoy, rather than separately at their respective offices.[25] One academician who feels guilt-driven in being away from colleagues believes that by telecommuting she must be more productive.[26] Another woman who enjoys telecommuting hopes that her observant children will follow her example and also want to do very good work.[27]

Yet another person who likes her job, but has personality conflicts with the supervisor, believes that working remotely enables her to maintain a distance from management and thus keep a job she needs. When Blue Cross/Blue Shield of South Carolina began its telecommuting program, the organization reported that workers' stress levels, fatigue, and low production problems vanished.[28]

Considering the advantages of telecommuting programs from a national perspective, specialists believe they will eventually help ease the city traffic problems that failing American transportation systems have caused and also provide the country with a 29 to 1 energy-use advantage over the automobile.[29] Indeed, 75 million barrels of gasoline imports could be saved if 14 percent of the work force telecommuted.[30]

As of 1985, the overall U.S. central-city traffic picture shows no improvement in traffic problems. According to the Census Bureau, a 64 percent majority of drivers continue to motor alone, and only 20 percent of riders carpool. While 6 percent of the population walks to work or takes a train or bus, a meager 2 percent works at home or uses other means of transportation.[31] Hartford, Connecticut, is an example of a typical metropolis that suffers from "gridlock" (hopeless traffic jam) now that its local highway system can no longer be expanded.

To answer the problem, the Hartford city government and downtown area companies are encouraging van-pooling (companies investing in fleets of commuter vans), employee flextime (working hours other than the regular 9-to-5 business hours), paying employees 50 percent of their bus-pass costs, and maintaining free parking facilities.[32] The traffic congestion problem could be eased if a larger percentage of Hartford downtown office employees telecommuted.

ELEMENTS OF SUCCESSFUL TELECOMMUTING PROGRAMS: TASKS, WORKERS, AND MANAGERS

Remote-work programs require not only well-defined and appropriate tasks and the best tools for accomplishing and transferring a completed job, but most importantly, dependable managers and workers who are properly trained for the program. The number of days that people telecommute each week is not critical to a successful remote-work program.[33]

Successful management of telecommuting operations in the increasingly information-based organization depends on having "objectives that are agreed upon in advance and clearly understood, self-discipline, and respect. This makes possible fast decisions, quick response, flexibility, and diversity," according to management systems authority Peter Drucker.[34]

Ultimately, however, the smoothly running remote program depends on whether it fulfills the expectations and fits the needs of potential teleworkers and on the way in which the workers are managed, claims Steve Shirley, president of F International Group, a software service company in which most employees—including managers—work from home.[35]

Tasks that are successfully managed in telecommuting programs are those where the individual already works alone handling information, such as in writing or editing, computer programming, word processing, and data entry. In general, these tasks require:[36]

- Little space and equipment.
- Little face-to-face contact.
- Long stretches of time when the worker can operate independently.
- Work that is project oriented, or where segments are clearly defined and produced within given periods of time.
- Communications or information transfer that can take place during the employee's office visits, by computer, direct cable, telephone lines, or via other electronic equipment.

Consequently, insurance companies, banks, computer firms, and software firms have been among the pioneer telecommuting users.[37] Remote work has also proved useful to stockbrokers, researchers, managers, people taking reservations, telemarketing employees, and trainers who design programs.

Financial analysts also rank among the 25 most popular remote-work categories. They profitably access information from marketing information databases, then use spreadsheet and word processing software to design and develop reports. The many other prime telecommuting occupations (ranked by weighted proportion of work that could be done remotely) include:[38]

Travel agent	Securities agent
Architect	Real estate agent
Writer	Computer systems analyst,
Word processor	programmer
Salesperson	Bookkeeper
Data entry clerk	Accountant
Insurance agent	Computer programmer

The ideal telecommuter is usually a highly motivated and skilled person who is committed to the organization and has volunteered for remote work. People who qualify as good telecommuters are known to be more productive at the office, more independent, less gregarious, and have both good organizing and time-management skills.[39] This does not necessarily mean that they are less sociable, however. A recent survey indicates that computer programmers, known to be good telecommuters and thought to prefer working alone, are actually very sociable individuals who particularly enjoy talking about their work and its problems.[40]

The successful manager of remote workers has a background in and is familiar with all aspects of telecommuting programs, has experience working remotely, is secure in using good supervisory skills, and confident in the telecommuter.

TYPICAL TELECOMMUTING APPLICATIONS

Decentralizing corporate offices is among the better-known telecommuting applications. For example, when Aetna Life and Casualty Company relocated a group of computer programmers from its main Hartford, Connecticut, offices to a suburb, they telecommuted daily with clients downtown via a teleconferencing system. Following this experience, the company expanded the use of teleconferencing to coordinate activities and train employees. Aetna is now developing a multisite system that spans the country.

Boeing also used teleconferencing to decentralize large groups of engineers working in diverse disciplines and in various locales around Seattle. By networking three or more conference rooms, groups worked together as committees during the developmental phases of the 757 aircraft.[41] Blue Cross/Blue Shield of South Carolina has decentralized the input part of its office operation beyond the office walls by allowing hospitals and 200 doctors' offices to file insurance information concerning patients' claims directly into the system.[42]

Teleconferencing systems used for telecommuting provide an excellent way for an organization to disseminate and collect information as well as to coordinate group activities. American Airlines uses a daily teleconference to brief most management levels coast to coast. Nuclear power plant operators around the world also access a computer teleconferencing system that enables them to check information on a daily basis and keeps them in immediate touch with each other.[43]

Telecommuting is a way of operating systems during nonbusiness hours. People who prefer to work evenings or at night can be of exceptional benefit to some companies. For example, those working for Aetna Life and Casualty in the early 1980s accomplished 40 percent to 60 percent more work because they experienced fewer interruptions and because the system ran faster. "Cottage coders" and "keyers" telecommuting for Blue Cross/Blue Shield of South Carolina transmitted their data via an automatic dialing arrangement so that information could move unattended.[44]

This has resulted in 1,200 medical-claim reports being processed each week. Workers are paid for each report processed, although errors are penalized by deductions taken from their salaries. Whereas the company saves considerably on fringe benefits, the workers have time home with their children and can set their own schedules. The telecommuting manager found that the only supervisory skill required was the need to be direct and consistent with her employees. Appearance and demeanor were unimportant to the job.[45]

Corporations can establish joint ventures by using teleconferencing systems. Such telecommuting systems allow dispersed units within a division as well as new acquisitions to jointly communicate, store, and manage electronic files. This can provide the company with a viable alternative to establishing new plants and thus give it a competitive edge.[46] For example, after Honeywell acquired a number of computer hardware, manufacturing, and software organizations, it depended on teleconferencing to coordinate the activity of specialists assigned to joint projects.

Teleconferencing is a well-known tool for internal corporate training needs as well as for postsecondary education and training. IBM and AT&T have developed extensive employee teleconferencing training programs, and the University of Wisconsin Extension Service offers many thousands of students courses in this way at more than 200 locations in the state. Separate networks deliver class material for sound (telephone) and slow-scan video transmission for pictures.[47]

Coauthor Tom Cross has used slow-scan video teleconferencing to teach graduate-level courses. These courses have included sessions where students call in from home, either as a group or individually. The results were dramatic in that students realized that a classroom was not required for effective learning to take place.

Telecommuting has found a home within the Fort Collins, Colorado, city government management system. The city fathers believe that telecommuting has given it a workable option for attracting and retaining certain kinds of employees, for managing the office space crunch, and for increasing employee productivity. In addition, city management has found that telecommuting improves communications between its employees and supervisors and reduces the amount of time government people spend in regular meetings.[48]

The Fort Collins telecommuters (currently 3 percent of the city's work force) are provided with the same tools at home that they are in the office, and these are linked to the city offices. This arrangement allows remote workers to use the same telephone features, office automation applications, and databases as their office counterparts.

By late 1985, the state of California is expected to integrate a telecommuting program among 200 of its 145,000 employees, according to state publications. Remote workers will include a variety of professions that range from medical doctors and attorneys to budget analysts. California will study the effectiveness of office size and costs, work centers, terminal or PC ownership, the effects of isolation on the employee, and the reaction of worker organizations. A planner with the California Department of General Services believes the state can experience a 20 percent productivity increase and pay for the project within 15 months.

Telecommuting is now an everyday part of auditing procedures in many companies. Auditors, accountants, and managers at the Houston office of Price Waterhouse carry portable personal computers to client offices to save "tedious hours of number crunching required to recalculate and update schedules," according to one of the firm's audit partners. They also take their computers home on weekends to use as a vital compiling and analyzing tool. When clients have systems that are compatible with their auditor's portable equipment, they can prepare data ahead of time on a diskette.[49]

Telecommuting has also become a way of transmitting banking information. The Continental Illinois National Bank and Trust Company has used telecommuters to transcribe recorded information and transmit data electronically to the bank's Chicago office computers. At Control Data Corporation, Minneapolis, 80 employees participate in a full- or part-time volunteer work program. By working from home or from a satellite work site, they save the company the expense of expanding its current facilities and help it save on growing heating costs.[50]

Employees can remain on call if they have telecommuting equipment at home. Apple Computer of Cupertino, California, allows employees to take home a computer that becomes their property after one year. During that time, employees are encouraged to use on-line office facilities to accomplish some work when they are home ill and to be available to answer questions about the job.[51]

Remote-work systems can provide a niche in the employment picture for the handicapped when they are well equipped, trained, and assisted by governmental or other agencies. Significant numbers of disabled people across the country have found employment within major business telecommuting programs. Lift, a Northbrook, Illinois, organization, provides a six-month training program for the disabled that is administered gratis via video and cassette tapes. Trainees are then hired as contract workers through Lift.

In the private sector, Control Data Corportion instituted a "Homework" program that allows 26 handicapped people to be gainfully employed. Working from such locations as Dayton, Ohio, and Hampton, Virginia, they connect their terminals (via telephone lines) to mainframe computers in Minnesota. While the employee benefits from "renewed self-esteem and productivity," the company benefits from worker expertise.[52] After establishing a program for the disabled, the company developed a second program, "Alternative Worksites," in 1979 for other employees. They are permitted to telecommute for periods of up to nine months on either a full- or part-time basis.[53]

Unfortunately, as evidenced by the following letter sent to coauthor Tom Cross, the well-trained handicapped individual has not always found a satisfactory work situation during the last years.[54]

July 30, 1984

Dear Mr. Cross,

I am a twice disabled homebound person. First I was completely paralyzed in childhood then had to relearn to do everything. Vocational rehabilitation put me through college, and I had almost 25 years experience as an engineer in industry when I was disabled the second time. I cashed in my life insurance to get this Radio Shack TRS-80 Model 2 because rehabilitation agencies wouldn't fund it or didn't have the money.

I have used computers as a tool to solve problems for almost 30 years and knew that some of this skill could be for hire to business/engineering clients. This was when I found out that it was easier to keep disabled persons on an endless education/training treadmill than it is to find them meaningful jobs. This is in the face of modern telecommuting and technology applications.

I found some strange political scandals behind all this facade of helping the disabled. Some of my more bitter disabled friends claim that less than $.01 of every dollar allocated for disabled causes ever reaches them in any needed help. . . .

Another strange political scandal is that almost all labor unions are against tele-commuting for the homebound. You will have to ask their leaders for mental gymnastics to justify this action. I am probably too mad about it to give an undis-torted view of their position. Since the Democrat party is heavily influenced by labor unions, the bias against telecommuting for the homebound shows up as in-ertia in the help the disabled get. That inertia shows up in businesses also when you try to sell them telecommuting skills on a contract basis—even if you demon-strate how you can save them $millions!!

The basic problem that keeps a disabled person from helping himself out of this situation seems to be [one of] a social and/or physical mobility which the home microcomputer with telephone-modem can solve. The problem gets worse then. I have found no clearinghouse where businesses can get telecommuters or where the telecommuters can get service contracts from businesses. Some of these business executives seem to like wasting $millions when a few telecom-muters will save it for them.

Have you or your readers got any solutions to this? I have even tried using salesmen on a commission basis with bad results.

Sincerely,

Kenneth Willoughby

In a search for telecommuters to handle an overflow of calls for ho-tel reservations in 1981, Best Western Hotel, Phoenix, Arizona, turned to the Arizona Center for Women (ACW). Following a two-week training period, some 21 to 53 minimum-security inmates were supplied with 21 terminals and telephones in the prison workroom. They keyed in data that was transmitted to the mainframe computer at Best Western head-quarters.[55]

Telecommuting provides a way for people to merge careers for new ventures. In one instance, a journalist and a social science specialist joined forces to establish an independent social research foundation.[56] Many people who are geographically separated research information, write books, and edit each other's material, and writers scattered from coast to coast send in their information or articles to newspapers and newsletter publishers.

Business affairs and teleconference training programs are also suc-cessfully managed remotely. In 1984, Virginia Ostendorf established a teleconference consulting and training company operated wholly via telecommunications.[57] Her company, Virginia A. Ostendorf, Inc., of Littleton, Colorado, manages its business affairs and all aspects of its tele-conferencing training programs through telecommuting. The 10 com-pany employees—writers, a training package designer, an expert on two-way, full-motion video transmission, and others—are dispersed from coast to coast. Clients and their students are located throughout the United States, Canada, and elsewhere.

During a training session, on-line students learn to accept and oper-ate all types of teleconferencing systems, from audio and audiographic to two-way, full-motion video systems. In a typical audioconference training session for Bell Northern Research in Ottawa, Canada, Ostendorf held a two-hour, on-line class for employees on "Effective Telephone Meeting." The students were located in Ottawa and Edmonton in Canada, Texas and North Carolina in the United States, and across the Atlantic in Maidenhead, England.

Before holding such a session, Ostendorf prepares and mails course materials to the students and determines a date and time that conveniently bridges the time zones of the various locations. Negotiations with clients, "classroom" preparations, the actual training sessions, billing, or payment do not require students or teachers to travel or meet personally.

There are any number of unique approaches to telecommuting. Steve Roberts, 32, for example, has made it his business to travel the country on a reclining bicycle while he writes articles for CompuServe's users' magazine. The bicycle is equipped with a Hewlett-Packard 110 portable personal computer, a solar panel to charge batteries, a citizens band radio, a tent, clothing, and other gear.[58]

The owner/manager of a 200-employee architecture firm telecommutes electronically with office staff and associates located anywhere from Tokyo to Sydney so that he can discuss "everything from equipment purchases to building permits." In contrast, George Chamberlain types computer programming instructions into a PC in his cell as part of a Minnesota State Prison rehabilitation program.[59]

OVERVIEW OF THE TELECOMMUTING COMMUNITY

Who telecommutes? Because there are many ways to work remotely, the telecommuter can be an employee or consultant on the payroll of one or more companies, a free-lancer who contracts for agency projects, or entrepreneurs who run businesses from home. Executives who carry their "electronic briefcases" home evenings and on weekends telecommute. Computer teleconferencing participants who transfer the results of their work to central files or to other persons telecommute.

The approximately 20,000 Americans who currently work at decentralized, suburban branch office operations and 30,000 who work at home part time can be considered telecommuters. The majority of these 50,000 people are in computer or related industries, although secretarial and clerical telecommuters work in all business areas.[60]

Electronic Services Unlimited (ESU), New York, reports that the average telecommuter is:[61]

- A male between 30 and the mid-40s.
- Five times more likely to be self-employed than a company employee.

- Producing better-quality work and more of it. (Supervisors agree overwhelmingly with this assessment.)
- Less distracted working at home than at the office.

A limited survey of telecommuters made by the New York University Graduate School of Business produced similar results but indicated that about half of the respondents work for other people or companies and only about 35 percent are self-employed.[62] The survey indicates that about 85 percent of the people who have a telecommuting option are satisfied with it and appreciate the convenience of being able to work at home; about 31 percent of them use telecommuting to earn extra money. The respondents appear to be productivity-conscious people who claim to spend one and a half workdays telecommuting in addition to their regular working hours. About 71 percent would choose to work part time at home if they had the option.

When the survey asked why working remotely appealed to the telecommuters, they said that productivity and working in one's own way ranked as first and second reasons, with earning extra money, saving on commuting time, and tax benefits following. The last reasons given for working remotely were that telecommuting was a way of easing conflicts and lowering overhead. While the respondents gave productivity as well as time with the family and for one's self as advantages, working too much and not having enough interaction (communications) with co-workers were given as disadvantages.

A 1985 survey designed to discover whether people in the Denver metropolitan area who owned computers were bringing office work home to process on home computers found that a high percentage of people were telecommuting part and full time:[63]

People Bringing Work Home to Process				People Remaining Home to Work			
	Sample	Yes	No		Sample	Yes	No
Total	85	51%	49%	Total	85	32%	68%
Male	43	51	49	Male	43	40	61
Female	42	50	50	Female	42	24	76
Ages:				Ages:			
18 to 24	13	69%	31%	18 to 24	13	46%	54%
25 to 34	26	65	35	25 to 34	26	35	65
35 to 44	24	38	62	35 to 44	24	17	83
45 to 54	15	33	67	45 to 54	15	27	73
55 to 64	6	50	50	55 to 64	6	67	33
Over 64	1	0	100	Over 64	1	0	100

A composite of surveys has produced the following information on employees work location preferences.[64]

Employee Preferences for Work Locations

	Year		
	1985	1990	2000
Go to office	40%	40%	30%
Split time between office and home	30	25	20
Work "on the road" anywhere	10	15	20
Work at neighborhood work center	10	10	15
Work at home	10	10	15

VARIATIONS IN EMPLOYEE ATTITUDES

As might be expected, employee attitudes toward telecommuting vary widely. There are devoted corporate employees who idealize the remote-work setting, volunteer enthusiastically for such an arrangement, but prove unable to organize their time and work without direct supervision. There also are upward-bound employees who immerse themselves in their telecommuting tasks, forcing their families to alter their lifestyles on behalf of the job. Most people experienced with telecommuting agree that employees have the need to socialize with fellow workers— even gossip—and to know they have not been forgotten by the organization or bypassed for promotion.

Occasionally employees are interested in using upgraded communications systems long before their companies are prepared to act. For example, Bob Allen, a business systems and records management analyst, recognized the benefits that his company, a brewery, could enjoy if the company implemented a flexible telecommunications program to mesh with its flextime policy.[65] So he studied his own work routine for ways to employ telecommuting techniques. His comments on telecommuting follow:

On working at home

More than 18 hours each week (over half the time actually worked) could have been done at home. Since we have flextime and I'm usually at work an hour or two after others have gone home, the company obviously trusts me to work on my own.

On conference calls

I already have the ability to dictate any work over my phone line . . . I can set up conference calling with up to seven extensions at once, and I would appreciate the opportunity to try it.

Since I started my study, I have started using conference calling. . . . The last two weeks I have told the party on the phone to hold while I get the third party on the phone. They have always agreed, and in every case so far we have reached a solution on the same call. I know this has saved me considerable time.

I also went to the brewery for meetings and then missed the party I went up to see. I could have solved the problem with a conference call. . . . I made a number of trips during my first week that I could have eliminated by using conference calling. Each trip to the brewery takes about half an hour. . . .

As a benefit to both the company and myself, if I were allowed the flexibility, I could even buy a communicating PC, charge it off my wife's business that she operates out of the home. The company could even check up on my work. Two of my employees live in south D_____ and have had some real difficulty getting to work during the winter weather. That plus sickness has forced them to stay at home. If we allowed/permitted/suggested/rewarded it, critical meetings, reports, etc., could possibly still be performed.

Sometimes organizations are ready for telecommuting and simply offer their workers the option. Jim Weidlein, a Boulder, Colorado, resident, telecommuted about 40 percent of his working time during the year and a half that he worked for a Denver teleconference consulting firm. Jim and most of the firm's other employees lived about 30 miles from Denver and communicated with the company via computer. By working remotely, they avoided battling the Denver rush-hour traffic and the city's pollution-choked atmosphere, worked in a quiet environment, and enjoyed flexible working hours.[66]

When asked whether telecommuting appealed to him, Jim answered, "I enjoyed the flexibility of my life and the variety of activities I could pursue. My working situation also allowed me to get additional personal things done in a more timely way."

Weidlein also believes that business can benefit from telecommuting programs because it can hire consultants located anywhere and communicate with them via computer. "This is more cost effective for business than using the telephone," he said, "and a more flexible arrangement for hiring and managing part-time help."

However, Weidlein advises managers considering a remote-work program to "ask if the firm is really served by using remote workers. The company should also ask if telecommuting will be more cost effective or productive than having the staff people in the office. In addition, managers should be certain that their personnel are sufficiently self-motivated."

Would Weidlein opt to telecommute again? "I would like to have telecommuting as an option for managing my work," he said, "but, personally, I found that I prefer being around other people during working hours."

TELECOMMUTING BARRIERS AND FAILURES

Most organizations still have misconceptions about telecommuting. They do not understand the remote-work process, its many benefits, or how to implement a telecommuting program. Many managers, in fact, still equate employee performance with punctuality, sociability, and appearance rather than the quality of work produced.[67]

About one half of the companies that have had telecommuting programs abandoned them within two years. An assessment of failed programs shows that it is usually the lack of standards and objectives in setting up projects that is at the heart of the problem. In addition, poor project management, inadequate programming standards and documentation for consultants, and a lack of communications between remote workers and the office are significant causes for failures when technology functions well and participants are carefully selected.[68]

Remote-work programs do sometimes run into technical difficulties. A four-year project carried out by Chicago's Continental Illinois National Bank was terminated in 1982 because equipment did not mesh and too many people were involved. In that project, secretarial work assignments had been farmed out and returned electronically.[69]

There can be problems in linking microcomputers to mainframes, although a number of software programs have been developed that virtually eliminate this problem. Present teleconferencing systems may not be sufficiently collaborative for groups that must work together. Furthermore, transmitting remote work can be slow and costly when analog telephone lines are used. As digital networks are put into place, work transmitted by phone will move much faster.

Because a telecommuting program depends on transfer of information, telecommuters will feel particularly frustrated when communications networks, office automation compilers, or telephone systems are down or there are hardware failures.[70] In an effort to overcome these and other problems, the city of Fort Collins, Colorado, performs preventative maintenance and computer file backup during those hours when the fewest workers will be affected.

In transmitting information from hotels, senders may not be able to directly access wall jacks and will thus be forced to use a less efficient device (acoustic coupler) to connect portable computers to telephones. Newer hotels offer "plug-ins," bypassing the need for couplers. In addition, the sender who depends on analog telephone lines found in most hotels may experience slow transmission speeds (costing more) and incompatible networks during transmission.[71]

Companies that integrate remote-work programs sometimes run into the simple logistical problems of getting office and telecommuting personnel together for face-to-face meetings. Or remote workers may have difficulty in reaching a busy supervisor or other office personnel.

Communications specialist Alvin Toffler suggests that telecommuting currently requires a stronger infrastructure, where machines have more intelligence and provide "less of a barrier between us and the machines." He also claims that computers are not currently personal enough and do only "essentially simple-minded functions." However, Toffler believes that computers teach us a different way of thinking about problems. By simulating or modeling problems, he believes computers help us examine and consider more alternatives.[72]

Certain companies fear bad publicity, such as that incurred by Blue Cross/Blue Shield of South Carolina or Washington, D.C., where tele-commuters who contracted to produce "piecework" saved the company a considerable amount of money but did not receive retirement benefits or raises over an extended period of time.[73] Wisconsin Physicians Ser-vice was also reported paying home workers who adjust claims 35 cents per hour less than their office workers doing the same job, and offered them neither health insurance nor retirement benefits.

Many businesses do not believe telecommuting is economically fea-sible for them. They have adequate working space, appropriate loca-tions, a sufficient number of personnel, and manage their other costs easily. Staff numbers remain static, office and traffic patterns are rear-ranged as needed, and the organization finds no advantage in having employees work flexible or nonbusiness hours.

Corporations sometimes face psychological barriers when it comes to telecommuting. Supervisors are uncomfortable handling remote, well-established work programs. For example, even though the Control Data Corporation program included 100 telecommuters working on a full-time or part-time basis (in 1984), upper management was less than enthusiastic about the arrangement. Comments were that bosses feared losing control of employees and productivity rates and that remote work made managing employees "more vexing than ever." In response, re-searcher Jack Nilles asks how a boss can determine if his employee is working even when he is at the office.[74]

Isolation is the primary psychological problem faced by telecom-muters. People who work remotely—like any group that is isolated for extended periods—need the stimulation of human interaction. Indeed, futurist John Naisbitt emphasizes that the more people are forced to live with technology, the more they want to be with people. There is little substitute for the stimulation, immediate feedback, and fun of direct contact in exchanging ideas.[75] Every organization that integrates a tele-commuting program will have to determine the amount of time that is appropriate for each employee to spend handling remote-work programs.

The following example illustrates this point: A technical writer in Salt Lake City worked in the Control Data Corporation "Homework" program for the handicapped. Although she had few problems during the eight-month telecommute period, she missed the camaraderie, politi-cal atmosphere, and office bureaucracy to the extent that she requested a return to the office. She also claimed that "to get in on promotions and raises, you better show up at the office." Similarly, a claim filer telecom-muting for Blue Cross/Blue Shield chose to return to the office when she realized that it was office socializing that broke up the tedium of her job.[76]

In addition, the families of people who work remotely sometimes lack respect for the telecommuter and intrude into his or her privacy with demands. Overeating and alcohol addiction can be significant problems for the telecommuter who finds the refrigerator or liquor cabinet altogether too inviting. Other psychological problems the telecommuter may face are discussed in Chapter 4.

DISBANDED TELECOMMUTING PROGRAMS

While it is easy for the manager to claim that telecommuting is not economically feasible or cost effective, those who cancel telecommuting programs generally have not adapted well to remote supervising where face-to-face interaction with the employee is limited. (See Chapters 3 and 4 for further discussion of failed programs.) Telecommuting specialist Gil Gordon suggests that a problem is sometimes that "telecommuting forces managers to use discipline."[77] Another specialist believes, however, that the basic problem with remote-work programs is that few people can "self-manage."[78] It also happens that telecommuting programs are often promoted by a single managerial influence. Consequently, when that manager leaves the company, the program can be quickly disbanded. Such was the case at the Federal Reserve Bank in Atlanta, Georgia.[79]

Business interest in telecommuting depends very much on local office space availability and other conditions which vary widely throughout the country and change from year to year. For example, Chicago, Denver, and Houston, all major cities, currently have relatively high vacancy rates in their central city areas.

This forced building owners to lower rents and developers to include various incentives within building structures—all of which discourage general local interest in remote work. The effect of other demographic trends, such as people returning to central city areas to renovate older property or suburban office parks being created on community outskirts, have yet to be determined in relation to telecommuting.[80]

GOVERNMENT REGULATORY ISSUES
AFFECTING TELECOMMUTING

Few localities have zoning ordinances that apply specifically to telecommuting. Most were written to preserve the exclusive character of residential neighborhoods and exclude occupations that create noise (car repair), unsightliness (advertising signs), odors (food preparation), and the like. Unfortunately, many ordinanaces were written so broadly that they have been interpreted to exclude remote work in certain localities, notably in Chicago and New York. Chicago now allows people to earn money at home as long as they have no employees.[81] When local government enforces such ordinances, it can assess heavy per-day fines if work is continued after a "cease and desist" notice has been issued.[82]

Ordinance problems are generally dealt with through enforcing performance standards or "criteria" rather than permitting or prohibiting specific occupations. The National Association of Home-Based Businesswomen (NAHB) has helped the telecommuter by categorizing restrictive ordinances and advising its members on how to overcome such barriers. Its approach is to describe the measurable, allowable effects of various occupational activities, thereby detailing their "performance criteria" within a model zoning ordinance.

Coralee Smith Kern of the National Association for the Cottage Industry takes a different position in regard to local ordinances. She believes that state laws and local zoning ordinance problems restrict a person's right to choose a workplace and that "many local (primarily city) ordinances prohibit residents from conducting any business out of their homes."[83]

In answer to the local ordinance problem, New Brighton, Minnesota, has written regulations that are combined with performance standards to cover work that is carried out in a multioccupational housing area. They are enforced upon complaint. The rules do not allow outside signs or storage, noise, vibration, smoke, dust, or on-street parking. Delaware (New Jersey) township also has a model ordinance that imposes specific standards covering the same areas. In addition, it demands that there are no changes to building exteriors, no increased traffic or water use, and that all occupations obtain a conditional use permit.[84]

Although land deed restrictions can also impede the telecommuting activities of residents, leasees, or subleasees, attorneys who review deeds should be able to catch the problem.

The presence of dish antennas can also pose problems for the telecommuter who depends on them for direct receipt of information. Most local areas and the FAA have restrictions regarding their placement. The city of Chicago, where antennas compete with cable television, has revoked permits for high-rise placement.[85] Because dish antenna placement helps determine the quality of reception, variances can be important to a remote worker.

Most communities require dishes to be placed in backyards and to be less than a specified height. Atlanta has a two-tiered approach, requiring a use permit but allowing dishes to be placed in the backyard as long as they are under 20 feet. Dishes that are placed on top of buildings are permitted "by right" and defined in the city code, though they must be approved.

UNCHARTED LEGAL AREAS

Companies that decide to integrate telecommuting programs may eventually face a number of legal problems. The problem is that there are as yet no legal reference books, cases, or statutes that apply to telecommuting. Workers' compensation insurance and work-for-hire agreements only appear to complicate the scope of employment doctrine, key to contract and tort liability. Employers would like to consider telecommuters as contracters, but the IRS says that companies must pay for workers who are fully salaried and work on a continuing basis using company-supplied equipment.[86]

Tax benefits, however, can accrue to the entrepreneur who works remotely when the home office is used exclusively for business on a regular basis. In such a case, deductions are allowed for office space, equipment repairs, data services, and other items. There may be a question as to whether computer teleconferencing or meeting customers only by phone can be considered to be holding a regular business meeting and entitles one to deduct home office expenses.

THE UNION POSITION

The AFL–CIO, Service Employees International Union (SEIU), and other unions have expressed concern that business will exploit the unorganized telecommuter who does electronic piecework just as it took advantage of garment and jewelry workers during the sweatshop era. At that time, home workers' wages were among the lowest, working conditions were unsafe, and employees were not offered fringe or pension benefits. In regard to these problems, the current telecommuting picture is quite mixed. Some companies pay relatively good wages for piecework but charge the worker for use of equipment and provide no benefits. Many others, in contrast, offer telecommuters the same wages and benefits as their office workers.

Karen Nussbaum, founder of 9 to 5 National Association of Working Women stated in 1983, "We think the opportunity for worker exploitation is rife with telecommuters," and "we think it (telecommuting) should be banned." Nussbaum also believes telecommuting benefits the professional rather than clerical workers whose diminished office presence may lose them a degree of bargaining power.[87]

The unions express special concern that business will exploit those people who are most vulnerable, such as women who must remain home with children, the elderly, and the handicapped. Unions also discern the following issues emerging for the telecommuter:

- Increased workloads.
- Reduced wages.
- "De-skilling" of workers.
- Equal pay for office and remote work.
- Promotion criteria for women.
- Training for more highly skilled jobs.

- Physical strain created by computer technology (visual fatigue, muscular strain, and general stress).
- The use of "electronic scabs" to break union influence.
- Work satisfaction.
- Job monitoring.

Thus far, telecommuters at the Equitable Life Assurance Society Office have complained that:[88]

- Pay is low, but there are demands for high output levels.
- Management shows little respect for workers.
- The company shifted "overnight" to automated office technology, allowing employees little or no participation in job design. (It is recognized that employee overinvolvement in implementing office technology can result in weak telecommuting projects, however.)

Major organizations are expected to lobby in 18 states for regulatory action on safety measures pertaining to the control of visual fatigue, muscle strain, and stress caused by using VDTs.

A U.S. Chamber of Commerce lawyer argues against the union position. Claiming that by prohibiting telecommuting and restricting the supply of labor in certain areas, the government would support unions in achieving a labor monopoly that eliminates competition for jobs and drives up union workers' wages.[89] Other telecommuting problem areas are discussed in the context of integrating remote-work programs.

Conclusion

The good news is that telecommuting is here to stay. The bad news is that it is not for every worker or every company. There have been probably as many failures as successes. For telecommuting to be productive, companies and employees will have to study the successful use of other technological innovations and understand the factors that contributed to it. Both companies and employees want their operations and careers to be successful. While this chapter provided an overview of the remote-work picture, the remainder of the book is devoted to making telecommuting possible, manageable, and livable.

NOTES

1. William Atkinson, *Working at Home; Is It for You?* (Homewood, Ill.: Dow Jones-Irwin, 1985), p. 10.

2. Hal Hellman, "Home Sweet Office," *High Technology*, February 1985, pp. 64–66.

3. Kathleen K. Wiegner and Ellen Paris, "A Job with a View," *Forbes,* September 12, 1983, pp. 143–50.

4. Neil Gluckin, "The Office Is Where the Workers Are," *Telecommunication Products + Technology,* June 1985, pp. 56–60.

5. "Viewpoint" (column), *Data Communications,* May 1984, p. 13.

6. Richard Slatta, Ph.D., "The Problems and Challenges of the Computer-Commuter," *LINK-UP,* June 1984, pp. 36–39.

7. Mike Lewis, "If You Worked Here, You'd Be Home Now," *Nation's Business,* April 1984, pp. 50–52.

8. "CW at NCC" (column), *Computerworld,* July 16, 1984, p. 28.

9. Eric R. Chabrow, "Telecommuting: Managing the Remote Workplace," *InformationWEEK,* April 15, 1985, p. 27.

10. Wiegner and Paris, "Job with View," pp. 143–50.

11. Lewis, "If You Worked Here," pp. 50–52.

12. "Newsfront" (column), *Data Communications,* May 1984, p. 48.

13. Lewis, "If You Worked Here," pp. 50–52, and Margrethe H. Olson, "Do You Telecommute?" *Datamation,* October 15, 1985, pp. 129–32.

14. Walter Kleeman, Ph.D.; Francis Duffy, Ph.D.; Michele K. Williams, I.B.D.; and Kirk P. Williams, I.B.D.; *Designing the Electronic Office: A Practical Guide,* to be published in 1986 by Van Nostrand Publishing Co.

15. Slatta, "Problems and Challenges," p. 38.

16. *Telecommuting Review: The Gordon Report,* vol. 2, no. 2, Gil Gordon Associates, Monmouth Junction, N.J., March 1, 1985, p. 2.

17. Chabrow, "Telecommuting," p. 28.

18. Edward Wakin, "Jobs a la Carte," *Today's Office,* September 1984, pp. 43–47.

19. Michael Antonoff, "The Push for Telecommuting," *Personal Computing,* July 1985, pp. 82–92.

20. Margrethe H. Olson, "Do You Telecommute?", *Datamation,* October 15, 1985, pp. 129–32.

21. Wiegner and Paris, "Job with View," pp. 143–50.

22. Antonoff, "Push for Telecommuting," pp. 82–92.

23. Gluckin, "The Office," pp. 56–60.

24. Lauren D'Attilo, "On the Job," *Datamation,* February 15, 1985, pp. 156–58.

25. Peter K. Dallow, *Telecommuting in Fort Collins: A Case Study,* October 15, 1985.

26. Antonoff, "Push for Telecommuting," pp. 82–92.

27. Verna Noel Jones, "Work Industry Sliding into Home Base," *Rocky Mountain News* (Denver), April 6, 1984, p. 51–W.

28. Chabrow, "Telecommuting," p. 28.

29. Slatta, "Problems and Challenges," pp. 36–39.

30. Atkinson, *Working at Home*, p. 12.

31. "Getting to Work" (Census Bureau chart), *The Wall Street Journal*, March 11, 1985, p. 25.

32. Daniel Machalba, "Like Other Cities, Hartford Has Gridlock; Unlike Others, It's Not Building Roads," *The Wall Street Journal*, April 29, 1985, p. 23.

33. Gluckin, "The Office," pp. 56–60.

34. Peter F. Drucker, "Playing in the Information-Based 'Orchestra,' " *The Wall Street Journal*, June 4, 1985, p. 28.

35. Steve Shirley, "Why Cottage Industries Fit the Information Age," *Management Technology*, February 1985, pp. 77–78.

36. Gerardine DeSanctis, "A Telecommuting Primer," *Datamation*, October 1983, pp. 214–20.

37. Hellman, "Home Sweet Office," pp. 64–66.

38. D'Attilo, "On the Job," pp. 156–58.

39. Cross Information Company.

40. Olson, "Do You Telecommute?" pp. 129–32.

41. Robert Johansen, *Teleconferencing Success Stories*, International Teleconferencing Association (ITCA), March 1985.

42. Eric R. Chabrow, "Telecommuting: Managing the Remote Workplace," *InformationWEEK*, April 15, 1985, p. 28.

43. Robert Johansen, *Teleconferencing Success Stories*, International Teleconferencing Association (ITCA), March 1985.

44. Hellman, "Home Sweet Office," pp. 64–66.

45. Margo Downing-Faircloth, "Would Working at Home Be Wise?", *Personal Computing*, May 1982, p. 42.

46. Johansen, *Teleconferencing*, p. 5.

47. Ibid., p. 11.

48. Dallow, *Telecommuting*.

49. Margaret Eisen, "Business Computing in the Home—How You Can Make It Happen," *Computer Dealer*, March 1984, pp. 62–68.

50. Downing-Faircloth, "Would Working," p. 42.

51. Ibid.

52. Ibid.

53. Atkinson, *Working at Home*, p. 21.

54. Kenneth Willoughby, letter; address: Box 317, Fairacres, New Mexico 88033; Compuserv ID# 71565,2005 for EMAIL.

55. Kathy Chin, "Home Is Where the Job Is," *InfoWorld*, April 23, 1984, pp. 30–36.

56. Anita Micossi, "The Ten-Second Commute," *PC World*, December 1984, p. 120.

57. Interview held with Virginia A. Ostendorf, Littleton, Colorado, October 14, 1985.

58. *Telecommuting Review*, March 1, 1985, pp. 1–18.

59. Wiegner and Paris, "Job with View," pp. 143–50.

60. Lewis, "If You Worked Here," pp. 50–52.

61. D'Attilo, "On the Job," pp. 156–58.

62. Olson, "Do You Telecommute?" pp. 129–32.

63. *The Denver Post/News Center 4 April (1985) Survey*, Talmey Associates, Boulder, Colorado.

64. "Employee Preference for Working" (chart); sources include: *The Wall Street Journal*, Honeywell Information Co., Cross Information Co., Bureau of Labor Statistics.

65. Bob Allen, "Trends in Information Management: IRM Long-Range Report," (a student's analysis of his job activities in a brewery), March 7, 1984.

66. Interview with James Weidlein, pres., Information Design, Boulder, Colorado, June 5, 1985.

67. Downing-Faircloth, "Would Working," p. 42.

68. Gluckin, "The Office," pp. 56–60.

69. Wiegner and Paris, "Job with View," pp. 143–50.

70. Dallow, *Telecommuting*.

71. Hellman, "Home Sweet Office," p. 65.

72. *Telecommuting Review*, March 1, 1985. p. 16.

73. David H. Rothman, "The Computer Cottage Industry Hysteria," *Washington Post*, July 7, 1985, sec. B, pp. 1–2.

74. Chin, "Home Is," pp. 30–36.

75. John Naisbitt, *Megatrends; Ten New Directions Transforming Our Lives*, (New York: Warner Books, 1982).

76. Chin, "Home Is," pp. 30–36.

77. Ibid.

78. Atkinson, *Working at Home*, p. 4.

79. Chabrow, "Telecommuting," p. 27.

80. *Telecommuting Review*, October 31, 1984, pp. 1–10.

81. *Telecommuting Review*, March 1, 1985, p. 8.

82. Tammara H. Wolfgram, "The Right to Choose Where You Work," *Profiles*, May 1985, pp. 38, 60.

83. Atkinson, *Working at Home*, p. 23.

84. Gregory Longhini, "Coping with High-Tech Headaches," *Planning*, March 1984, pp. 28–32.

85. Ibid.

86. Wiegner and Paris, "Job with View," pp. 143–50.

87. Ibid.

88. David Farkas, "White Collars with Union Labels," *Modern Office Technology*, May 1985, pp. 118–22.

89. Lewis, "If You Worked Here," pp. 50–52.

2

Driving Forces behind Telecommuting

Just as technical and economic factors are fueling the development of tele-commuting, so are:

- Current general economic conditions.
- Heightened global competition.
- Evolving business management styles.
- Shifting societal, work, and lifestyle patterns.
- New demographic trends.

In addition, business functions that are normally carried out by people are being replaced by computerized systems that support airline or hotel reservation staffs, claims adjusters, and many other office workers.

This chapter explains the effect that evolving technology and economic, social, and demographic trends have on telecommuting. Finally, it discusses those areas where telecommuting trends are not yet apparent.

EVOLVING TECHNOLOGY

It is primarily the vast array of sophisticated new communications and computerized equipment—available at decreasing cost—that is spurring the growth of telecommuting. With these faster, more dependable, portable, and "friendly" devices, the business world is able to extend the communications networks of its automated offices to the telecommuter at home or elsewhere.

As the cost of micro or personal computers (PCs) has decreased, they have become enormously popular in business and in the home. In a business setting, the PC connected to a local area network (LAN) or telephone line enables office and remote workers to send messages or transmit work back and forth and use file servers (electronic file cabinets), printers, and other peripheral devices. Even in instances where PCs are used only for accessing databases and transmitting electronic mail within the corporate setting, they are often still considered cost justifiable.

PCs are becoming increasingly valuable for remote work as manufacturers enhance such features as memory, hard disk storage, graphic displays, PC-based video teleconferencing, and applications software. Applications software can provide financial analysis as well as job-specific and industry-specific programs. As a popular home device, the PC also enables the majority of owners to conduct at least some business at home even when they are employed full time elsewhere. This is reflected in the 50 percent increase in the number of people giving home addresses as business addresses during the last 10 years.[1]

Small, portable PCs, now produced with a growing number of features, have a special influence on telecommuting development. Because many models fit easily into a briefcase, people are able to leave noisy, congested city areas early to work evening hours at home. Portables also permit the occasional off-site worker to become accustomed to telecommuting.

Dr. Nicholas Negroponte of the Massachusetts Institute of Technology computer labs uses the Radio Shack Model 100 effectively when he travels. He fits rubber "mufflike" acoustical modem connectors over the handset of his portable, connects the portable to a telephone receiver, and dials his electronic mailbox number. Using the computer keyboard, Negroponte then orders the system to send his messages to the portable's memory banks. When he has time available, in flight or elsewhere, Negroponte has the portable print out his messages so that he can answer them.[2]

Many executives who are busy managing people and taking telephone calls during their regular work hours become effective telecommuters once they have carried their portable computers home. The tendency is to leave them at home for work that must be done without distraction. When people discover the value of using a computer at home or in the office, they frequently purchase a second one for the other location.

Because most employers do not buy computers for employees who are not regular telecommuters, executives of banks, brokerage houses, insurance companies, and similar service companies who must sometimes work at home often purchase computers to do so. They then approach their companies to secure remittance for the purchase.[3]

Aware that employees generally want to have the same type of computer at home as they have in the office (but with additional educational and recreational software programs), vendors frequently grant discounts and special services for employees of companies that purchase a certain quarterly volume of equipment. A Dallas, Texas, computer sales manager claims that some organizations are giving computers to employees as annual bonuses rather than cash. These companies believe that workers perceive the computer as having greater value and may use the home computer to do extra work.[4]

Although evolving teleconferencing and electronic mail systems represent fairly simple communications software technology, they are dramatically affecting the way many corporations manage their information flow and work. In addition, continually upgraded computer hardware and software, satellites, local area digital technology, and cellular radio systems are also contributing to the growth and development of the telecommuting area.

Software programs such as PHOTOBASE™ allow users to merge or integrate pictures or graphics with text from a database system. Images from a video camera, videocassette recorder, or laser disk are fed into a PC–EYE™ printed circuit board where they are digitized and treated as any other file.[5] Graphic representations that range from photos and drawings to blueprints, signatures, and logos can all be handled in this manner. Systems that interact with computers in this way enjoy an advantage over those (such as facsimile systems) that operate independently.

Voice mail systems are being rapidly incorporated into the automated office picture, and "talkwriters" (voice-controlled computers) that enable executives to dictate letters directly into them are currently being perfected by major computer companies. Most systems now in use compare a speaker's sound wave patterns with those the user has previously stored.

At present, this type of equipment is overly sensitive to background noise and fails to recognize more than one voice at a time. However, systems are being perfected that will recognize up to 2,000 words and enable people to call up files, sales charts, and inventories and handle fund transfers for banks and stock brokerages.[6] See Chapter 7 for more information on voice mail.

Refinements in touch-screen devices and equipment that reads handwritten numbers and letters are also expected to bolster telecommuting. In this latter instance, the device converts writing to digital data that appears as information on the user's screen, eliminating the need to key in written information. In its current stage of development, this equipment is considered costly for conventional, everyday use.[7]

New devices and specialized systems permit a number of secretarial services to be performed remotely. For example, one system feeds taped dictation or information to a home operator via the telephone system. The telecommuter, using headphones and a stop-and-go pedal, then transcribes that material. When the transcription is completed, the work is sent via modem and telephone line to an office printer.

Communicating copiers will soon be able to interact with document distribution systems and the electronic file cabinet. And on the home front, "digital television" that will replace older TV sets is expected to provide purchasers with complete computer services that will enable them to telecommute.

Telecommuting technology is also being enhanced as major equipment producers and communications networks apply universal standards to their products. These standards have been developed by international organizations so that most office devices will eventually interact and their users can communicate with one another.

In addition, national regulatory legislation is having a positive influence on the telecommuting picture. For example, the presence of new telecommunications carriers is increasing the user's choice of networking alternatives. This is expected to reduce the cost of using computers to send remote work and electronic mail over long distances, for managing home banking, shopping, and for other purposes.[8]

ECONOMIC ISSUES SPUR TELECOMMUTING

A number of economic factors are forcing business organizations to evaluate newer technology and consider using alternative operating methods, including telecommuting. Primary among those factors is the globalization of economic markets, a trend that has increased U.S. and worldwide competition, creating pressure on business to be more productive and make timelier, quality decisions. At the same time, globalization has put organizations at the mercy of worldwide market fluctuations which are beyond their control.

Telecommuting is also being influenced by the declining cost of computer technology relative to the inflationary costs of running an organization. Such mounting organizational costs include: personnel wages and benefits, relocation of skilled people, communications, travel, property, energy and utilities, and others.

WORK MOVES BEYOND THE CORPORATE WALLS

Many companies no longer need to locate their offices in traditional places. Telecommuting techniques provide them with the means for dispersing workstations throughout building areas to other facilities, work centers, and employees' homes. The "intelligent" or "smart" building emerging from coast to coast can accommodate such offices by offering a variety of services and/or space, temporary or shared facilities, or a new branch office for companies.

Intelligent building developers claim their services can cut tenant costs by 20 percent or more and are cheaper than they can be purchased piecemeal. Developers also contend that building landlords will stand to gain a marketing edge and profit by offering these services in various ways.[9] It should be noted that a building's "intelligence" is built into its floors, ceilings, walls, and communications system and cannot be discerned by the man on the street.

The evolving intelligent building will eventually control its own internal heating, lighting, air conditioning, and other systems, as well as integrate automated office and communications devices within its structure, connecting them to public and private outside networks. It will thus give tenants access to a growing number of specialized telephone services such as call forwarding, handle their data transmissions and computer and voice messaging. See the Resource Guide at the end of the book for a list of intelligent building systems and services.

Although critics of the intelligent building trend consider it to be only a building developer's marketing ploy, integrated building construction offers the modern office many advantages. Among the better-known intelligent buildings is the Renaissance Centre located near Dulles Airport in Virginia. It offers all of the above-listed services and enables users to access databases. More information on intelligent buildings can be found in Chapter 5.

The electronically outfitted work center also serves as a decentralized office. Such centers are currently located in unused schools and other buildings, day-care and shopping facilities. United Technologies Building Systems Worldwide Business Centers has established commercial work centers that operate in 21 major cities and provide sophisticated offices that can be rented by the day or year. In addition, the system offers clients translation service in foreign countries. Hotels and airlines are following suit by installing business centers with a broad range of equipment and services throughout the world.[10]

Headquarters Company (also referred to as HQ) and Omnioffices, located in Atlanta and nine other cities, are companies that also offer work centers. They supply fully equipped offices in shared-service buildings. Omnioffices' full-service plan provides mail, phone, private office, access to conference rooms, and other shared services including the free use of 14 sites in different cities.[11]

Orlando 4000, a "computer condominium" located in Florida, provides companies with alternative in-house data-center capabilities that can be purchased in increments. The subscriber chooses from a variety of systems and software that manage required tasks. Among the features offered are disk and tape storage, communications, and security. Communications are provided through Martin Marietta's network nodes in 11 major U.S. cities and London.[12]

Orlando 4000 has a human information system available 24 hours each day, seven days a week through a toll-free hotline. This provides clients with professional help in systems operation and tuning, product support, database management, and other pertinent areas. Thus the user has experts available in critical areas of information processing technology without incurring the full expense of hiring such people outright.

The Orlando 4000 concept, arranged on a contracted, fixed-price monthly rate, is expected to benefit some companies more than time-sharing systems where costs vary widely from month to month. It is also expected to help data processing managers whose duties leave little time for managing information. The financial benefits are reduced costs and improved cash flow.[13]

The commercial "smart" home moves employees another step away from the decentralized office by enabling residents to work where they live. Smart homes utilize computer technology to both control the home environment and provide residents with a full computer system.

One smart home system uses radio and power line carrier technology, runs on a wireless system, and is managed by either an Apple® or IBM PC® compatible computer. (A listing of such systems is included in the Resource Guide.) Residents draw a floor plan of the house, then program the system so that it will set the thermostat, control appliances, respond to fire, water the lawn at a given time, etc. Once the system is programmed, the computer can be disconnected and used for other purposes (communications, home banking, etc.) while it continues to manage the home environment.[14]

Smart housing developments in California that are currently being built near high-tech centers as telecommuter residences are controlled by centralized computer facilities.[15] For example, the Eaglecrest development in Foresthill, California, is expected to:[16]

- Eventually have 360 homes with data communication links for telecommuters.
- Cost $140,000 to $190,000 per unit.
- Receive remote work via a large satellite dish.
- Include computer-run home management (controlling temperature, lights, etc.), the use of up to 12 phone lines, and a set of preselected software packages. An educational computer will be tied in to local schools.
- Cater to writers, engineers, computer programmers, and other creative professionals as well as small entrepreneurial enterprises.

A neighborhood center offers copying machines and office services. Some New York City condominium developers now include computers as normal building features as well as offer subscriptions to various data banks that provide stock quotes, airline schedules, and other services.

Just as smart office buildings offer centralized information support services for tenants, certain cities are attempting to expand their economic bases by developing advanced telecommunications infrastructures for local companies and others that they hope to attract to the area. This requires such cities to have urban planners assess the usefulness of the community's existing systems, then consider which services to offer, the regulation and use of those services, the investment in, establishment, and promotion of those services.[17]

Under the auspices of AT&T and CBS Inc., a community of 200 homes in Ridgewood, New Jersey, has been computerized. Some homes use stand-alone devices while others use equipment that connects to a TV set where the screen works in conjunction with the computer.[18]

Forest City, Iowa, population 4,350, is on its way to becoming an ideal telecommuting environment and one where remote work is particularly useful to the population during long Midwest winter months. Not only do computers run the town's businesses, but a venture developed and supported by Control Data Corporation of Minneapolis, the Winnebago Company, its founder and the foundations he supports, a local Lutheran college, and the city's schools have established computer learning centers within the Winnebago Company facility, city schools, and the college.[19]

Even airplanes in flight can be workplaces now. Most major airlines have installed "Airfones" for in-flight telephoning and permit passengers to use portable and pocket computer terminals. In addition, certain airlines plan to rent portable terminals to passengers. Similarly, private aircraft are being outfitted with communications devices and are serving as complete offices that can be rented.[20]

EVOLVING ORGANIZATIONAL STYLES

The importance of an organization's information flow for management cannot be overemphasized. And, as we noted earlier, telecommuting enhances that flow. "Information serves as the axis and the central structural support" of the organization of the future, quite apart from technology, according to management theorist Peter F. Drucker. As part of that picture, he contends, business now requires employees who ask "who needs what information, when, and where?"

Successful information-based organizations share a number of communications characteristics. They rest on a circular information flow pattern where communications move "from the bottom up and then down again." In contrast, in traditionally run companies, information generally flows from top authority down. Citibank, some Japanese trading companies, and Massey-Ferguson, the Canadian tractor maker, are currently redesigning their managerial structures around the circular flow of information.[21]

Information-based companies also prefer decentralized decision making, manage their employees in a flexible and participative manner, and create a sense of mission among them. As a result, employees who work for such companies appear to be more productive, have less absenteeism, and a greater sense of well-being.

Ideally, responsible individuals working for an information-based company ask what performance and contribution the company expects of them.[22] Psychologists widely acknowledge that people who are brought into the decision-making process and granted a degree of autonomy will go out of their way to meet company goals.

Major companies like IBM and General Motors Corporation that are facing tough foreign competition are following the new organizational trends. They find that by decentralizing managerial responsibility, moving decision making down the ranks, trimming their staffs, and eliminating unnecessary paperwork, their staffs are making better, faster decisions. The Dana Corporation of Toledo, Ohio, a car- and truck-parts manufacturer, followed this formula. Whereas it was once a billion-dollar corporation employing 600 people, it has trimmed its staff to 85 people and increased its earnings to $3.6 billion.[23]

When youth has the choice of working for traditional or new-style corporations, it is likely to choose employment where workers partake in ownership—whether psychic, real, or both—and where emphasis is placed on personal development, according to futurists John Naisbitt and Patricia Aburdene.[24]

CHANGING ROLES OF MANAGEMENT AND TECHNOLOGY

The manager in today's rapidly changing office environment is faced with an increasing number of technological and personnel issues. New organizational and management styles will be required to handle these issues before telecommuting becomes widespread in today's homes, work centers, or office building settings. This section focuses on the changes in business operations and management that support telecommuting in the new environment.

The Computer Influences Organization Hierarchy

Computerized technology is creating a number of major structural changes in traditionally organized "smokestack" organizations. One change is to reduce the number of management levels between corporate presidents and line workers in large organizations. Several factors influence this reduction:

1. The computer permits business to respond faster and more effectively to clients' needs. Before the days of automated office technology, upper-level decision makers in large, authoritarian organizations tended to receive delayed, distorted, and inaccurate product/market information after it passed through many management levels.

2. The computer facilitates business information flow by allowing all managerial levels to access it easily and without intervening layers of personnel handling or setting it aside.
3. The computer enables managers and telecommuters in different parts of a plant, or in company offices dispersed throughout an organization, to access information and communicate via electronic mail. Thus, the manager does not need to be at any particular location—inside or outside of corporate walls—to communicate or receive required information. As a result of the widespread access to necessary on-line information, management perspective has shifted one step beyond the traditional, centralized office.

Independent Business Units (IBUs) Emerge

Management's new time/place perspective has produced a number of business trends. The trend most relevant to telecommuting is the development of small, independent business operating units (IBUs). IBUs can comprise a single person or a large group and can be located at the corporate office or almost anywhere else. In fact, they need not be located in any physical proximity to one another. These operating units generally comprise professional or information workers such as engineers, clerks, floor managers, lawyers, and other worker categories.

Electronic communications technology allows IBUs to work on any type of task force, ranging from an Equal Employment Opportunity Commission (EEOC) to committees concerned with toxic waste, benefits, new products, sales, or education. IBU personnel may work from home, from the office, on the road, or at a resort.

Communications Becomes Key

The subtle but important point here is that groups within the organization are able to exchange information and work with one another—in effect, to telecommute. As shown in Table 2–1, managers spend an enormous amount of time, in fact almost all of it, communicating. Hence, they are "communications intensive" in nearly every aspect of their functioning.

Furthermore, as the numbers of decentralized information workers at distant locations grow (internationally and across many time zones), the need for effective, efficient technology and management leadership becomes imperative. This means that managers will spend more time guiding, coordinating, motivating, and educating their employees than managing in the traditional sense. It should be noted that professionals require more interaction with supervisors than other employee levels.

TABLE 2-1

Present Manager's Distribution of Time	
• Communicating information transactions	
Meetings	30%
Telephone	20
Travel	20
• Seeking information transactions	
Desk work	30

Future Manager's Distribution of Time	
• Communications/information transactions	
People interfaces	40
Meetings	
Presentations	
Audio conferencing	
Video conferencing	
• Travel	10
• Seeking information transactions/	
System interfaces	50
Dictation	
Telephone/voice mailbox	
Computer conferencing	
Viewdata—data systems	
Decision support systems—assisted "thinking"	
Computer-assisted retrieval	

Present Nonmanager	
• Inputting information	
Dictation, typing, data entry, and proofreading	20
• Interpreting information	
Telephone, mail	35
Away from desk	30
Waiting for work	10
Absent	5

Future Nonmanager	
• Information analysis	10
• Information transactions	50
Project research	
Meeting coordination	
Arranging travel	
Budget tracking	
Purchase order tracking	
Researching/conferencing	
• Administration coordination/support for meetings	
Telephone, mail	35
Absent	5

SOURCE: Cross Information Company.

New Management Structures Emerge

The manager's role will be more to direct the flow of information and employee interaction (according to company goals) than to control the interactions within the IBUs that telecommute. This distinction grows from the awareness that information flow in an organization cannot be controlled, but must be guided. This has led to the development of "virtual management" (VM). See Figure 2–1.

FIGURE 2–1

Vice President—Technology Marketing

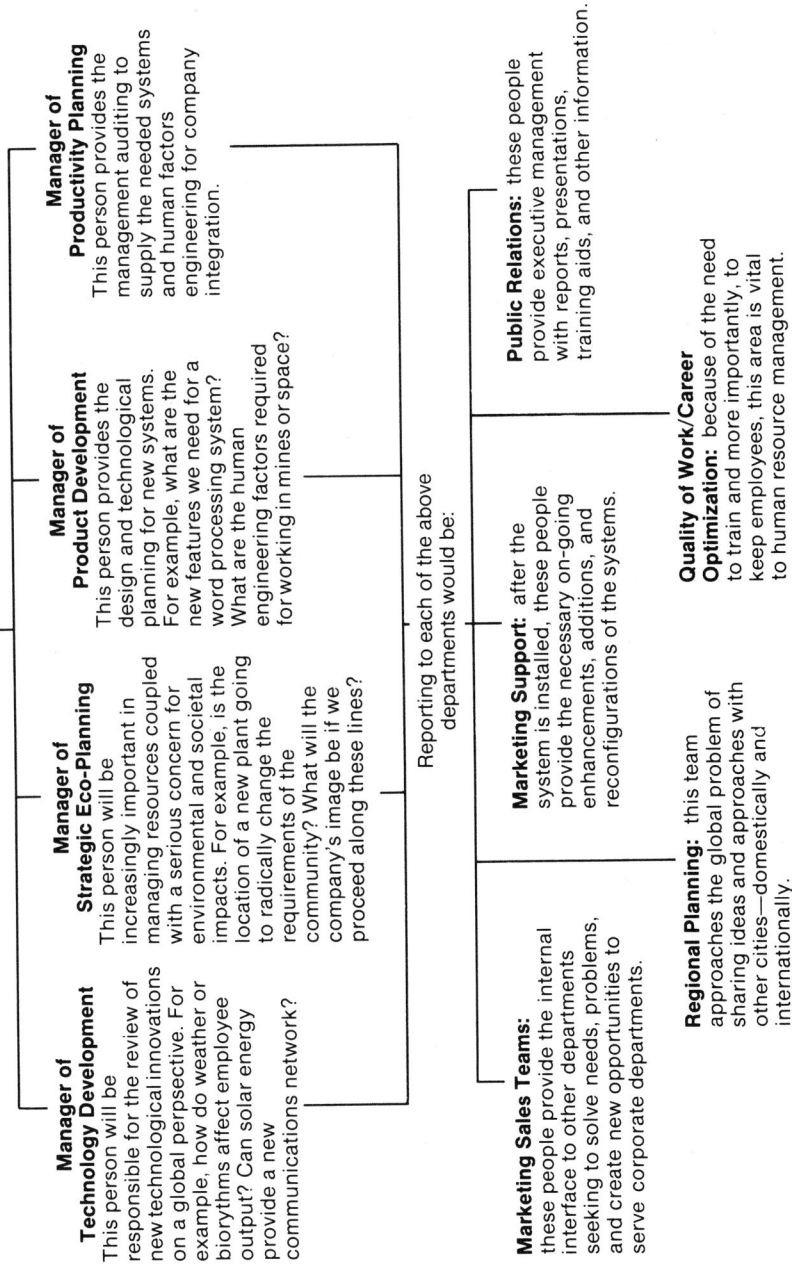

Manager of Technology Development

This person will be responsible for the review of new technological innovations on a global perpsective. For example, how do weather or biorythms affect employee output? Can solar energy provide a new communications network?

Manager of Strategic Eco-Planning

This person will be increasingly important in managing resources coupled with a serious concern for environmental and societal impacts. For example, is the location of a new plant going to radically change the requirements of the community? What will the company's image be if we proceed along these lines?

Manager of Product Development

This person provides the design and technological planning for new systems. For example, what are the new features we need for a word processing system? What are the human engineering factors required for working in mines or space?

Manager of Productivity Planning

This person provides the management auditing to supply the needed systems and human factors engineering for company integration.

Reporting to each of the above departments would be:

Marketing Sales Teams: these people provide the internal interface to other departments seeking to solve needs, problems, and create new opportunities to serve corporate departments.

Marketing Support: after the system is installed, these people provide the necessary on-going enhancements, additions, and reconfigurations of the systems.

Public Relations: these people provide executive management with reports, presentations, training aids, and other information.

Regional Planning: this team approaches the global problem of sharing ideas and approaches with other cities—domestically and internationally.

Quality of Work/Career Optimization: because of the need to train and more importantly, to keep employees, this area is vital to human resource management.

Courtesy Cross Information Company.

VM is an information-based organizational process that has three major components:

1. VM adapts to technological and societal changes by allowing management to guide employees, regardless of the geographical distances or the time differences between them.
2. VM promotes the use of technology as a supplemental, integral part of the management process, but not as a replacement for it.
3. VM seeks to facilitate the development of all information-based technologies throughout the corporate enterprise, recognizing that companies too often fail in their goals to produce goods or increase performance because they do not manage information adequately.

Just as VM recognizes that most of today's business is knowledge based, it also recognizes that while information may be contained in computers, it is found predominantly within "the heads" of company employees. And as workers are dispersed in remote-work settings, their need to access information is imperative. Gathering information from people separated by distance or time is one of management's challenges in telecommuting programs. Management and technology can facilitate, inhibit, or make this process impossible.

Managing human factors is one of the most significant aspects of the virtual management process but also one of the gravest barriers to telecommuting. Society, often diagnosed as being information rich but interaction poor, has developed an increasing resistance to technological cure-alls. However, priorities are currently being reversed and such human environmental needs as ergonomically sound furniture, terminals, lighting, heating, etc., are becoming an important part of office systems design. Custom "designer" offices that are now in vogue are a product of this general concern.

NEW MANAGERIAL POSITIONS

New types of managerial positions are evolving to support technology and the flow of information in business. For example, the position of information broker has emerged to handle the function of information gathering. "The approach to seeking information in the work setting will drastically change by the late 80s," according to Rudy Munguia, vice president of Intelligent Buildings Corporation™, Boulder, Colorado. He believes that

> There will soon be so many on-line resources that the worker will be frustrated and require an "information broker" to determine the best of many pathways to take in finding information. This individual will act as a consulting middleman who operates as a clearinghouse. He or she will find the appropriate sources, then "plug" the worker into them.

Munguia also believes that whereas workers today are "task" driven, those of the future will be "solution" driven and need more guidance in finding their answers.

Currently there are only 125 information brokers listed in the country, according to Norman Goode, publisher of *Micro Moonlighter Newsletter,* Lewisville, Texas. At present, they are "paid researchers with access to databases, who submit customized reports on specialized subjects to business clients," and provide business with "a private librarian at its beck and call," said Goode.[25]

Not all management analysts consider telecommuting necessary or advisable for business organizations. Harvard faculty members specializing in social and organizational psychology and industrial and environmental design believe that the willingness of organizations to accept telecommuting reflects their failure to support the vital business objectives of employee "participation, collaboration, and privacy" within the office environment.

They contend that companies should provide employees with a variety of "activity settings" (work areas) that support many different and often apparently contradictory tasks. Activity settings focus on the shared needs of work teams (ad hoc groups, quality control circles, and task forces) and the whole office work force. They also contend that while employees need "home bases" (private offices within company walls) that give them a feeling of stability in the organization, the all-purpose workstation for each person is inadequate for this need.[26]

New technology and communications techniques permit companies to study and assess their marketing positions and client base so they can consider selling services that are built around that technology. For example, after Control Data Corporation reassessed its marketing position, the company determined to develop Ticketron rather than only sell tickets and vending machines. Thus, Control Data currently sells "the entire solution instead of only part of it."[27]

EMPLOYEES BECOME FREE-LANCERS AND FORM THEIR OWN PROFESSIONAL CORPORATIONS

The advantages of operating independently are quickly apparent to people who work remotely as either corporate employees or as contracted labor for an agency. While the corporate employee is required to do assigned work and must abide by company agreements and regulations, the entrepreneur can—at least theoretically—choose his or her work, establish desired guidelines, and negotiate acceptable agreements. And, whereas the corporate employee often suffers tedium in doing repetitious remote work, the entrepreneur must handle the variety of chores required to run any business, such as doing accounting, purchasing supplies, etc.

Of course, there is always risk in establishing a business dependent on remote work. Equipment must be purchased, work promoted and contracted for, etc. The principal advantage in managing one's own remote-work business is that there is low overhead, certain tax breaks are available, and there are financial rewards when the business succeeds. Flexibility of hours, less need to commute, and other factors remain much the same.

Statistics show that while 58 percent of free-lance telecommuters earned more than $5,000 in 1977, only 8.9 percent of them made more than $100,000. One successful entrepreneur says that while her work went well and she was home "with" her children, she was too busy to really be there "for" them.[28]

Although the corporate or agency employee and entrepreneur approach work in different ways, both must:

- Be alert to the laws regarding working at home in their respective localities.
- See that their home offices are appropriately and comfortably outfitted.
- Educate their families and friends to respect their need for uninterrupted working hours.
- Develop a new kind of relationship with employers or clients.
- Organize their working time so they will not feel isolated.

NEW ATTITUDES TOWARD SCHEDULES AND CONTRACTS

Changes in organizational operating style often require adopting new work time arrangements. "Changes are shaking the twin pillars of the tradition-bound world of work, time and place," and are now involving one in five American jobholders, according to the Work in America Institute, Scarsdale, New York.[29] Indeed, the government estimates that more than 10 million full-time employees currently work flexible schedules or compressed work weeks.[30] Thus, "flextime" (and similar names) has become a consideration in most companies across the country and is part of the telecommuting scene.

Not only is flextime a convenience for employer and employee alike, it also promotes greater productivity. A two-year study of state government computer programmers showed that overall worker output increased by 43 percent when employees shared the same equipment. In contrast, the production of a control group working traditional hours increased by only 5 percent.[31]

The concept of selling or contracting one's work rather than time is also growing in popularity and is expected to affect remote work. Companies that use telecommuters are advised to set up and supervise remote-work tasks on a per-project basis where clearly defined segments can be handled during relatively short periods of time.

"Leasing" or "contracting" temporary help to work on a per-project basis is often encouraged when an organization needs to expand or limit its operations. This is sometimes referred to as "flex-staffing" or "lean-staffing" and is considered a responsible business approach.[32] The following benefits for organizations that use temporary help are considered to be:[33]

- Overhead is reduced.
- Record-keeping on leased employees is virtually eliminated.
- Federal paperwork and tax forms are eliminated because leased personnel are not technically employees of the company that uses their services.
- Fringe benefits are not required.

F International is a British agency that dispatches its worldwide staff of 1,000 persons to temporary assignments according to their expertise.[34] Its employees, 90 percent of whom are women, are versatile in their respective areas. Hardware/software evaluation, conversion of systems, project management, maintenance, and environmental analysis are among their technical capabilities. In addition, the staff is able to assess a company's security considerations, select jobs and managers, and train staff personnel. The organization believes that 20 to 30 percent of most large firms could have telecommuters. Few other temporary agencies have begun supplying organizations with telecommuters for flex- or lean-staff programs.

DEMOGRAPHIC TRENDS SHAPE TELECOMMUTING

Many demographic trends are influencing the telecommuting picture. For example, while the number of people available for jobs is tapering off, a growing shortage of workers is developing in various sectors of the economy. In addition, a growing part of the population (currently 25 percent) is unwilling to relocate where jobs are known to be available.[35]

Among executives, more than 42 percent are reluctant to move, and almost 7 percent are against any move.[36] This may reflect inflationary relocation, housing, and mortgage money costs, as well as less general willingness to incur family upheaval for the sake of job change. This is particularly true when a spouse is forced to change employment along with the primary earner.

As the shape and size of American families are changing, the number of two-parent families continues to plummet.[37] Among these changes, the Yankee Group research and consulting company believes a "Taffie" (technologically advanced family) is emerging that is more involved and comfortable with technology than other families.[38] The Yankee Group considers the way in which these families use electronic equipment to be a "distant early warning system" that may signal the success or failure of those electronic equipment categories.

Among their characteristics, Taffie mates tend to be college gradu-
ates between the ages of 25 and 50 who both work and have children under 18. Their incomes are generally over $25,000 each year, they own the
phones in their homes, and they often subscribe to such non-AT&T services as MCI and Sprint. Furthermore, Taffies often own such devices as
electronic typewriters, answering machines, and compact disk players
and subscribe to such services as The Source,™ CompuServe,™ or Dow
Jones News Retrieval®.

Current demographic studies reflect a growing number of women in
the U.S. labor force:[39]

Year	Percentage of Women Working
1970	38%
1980	42
1990	46 (expected 125 million total)

Female working patterns are also changing. For example, women of
child-bearing age are taking only a leave of absence when they have children rather then retiring from their positions. And they frequently become a permanent part of the dual-income family as they take on more
decision-making positions.[40]

As a consequence, the U.S. Department of Labor has been studying
the impact of computer work on the 13 million women who hold 80 percent of all administrative support positions in the country. It also plans
to determine whether computer use is to women's advantage.[41]

Along with the growing percentages of women in the workplace,
women's business roles are changing:[42]

- Women now own about 3 million business enterprises.
- One woman of every four is a sole proprietor.
- More women are breaking new ground as bosses.
- Women have gone into business on their own "at a rate six times
  faster than men did from 1974 to 1984."[43]

Women's attitudes also appear to be changing. Not only are women
becoming receptive to technology, but, as studies by a major advertising
agency reflect, they have sharply increased career ambitions and a need
for social supports such as health clubs and "social surrogates." These
are defined as "things that allow a person who is alone to feel that he
isn't," such as video games and computers.[44]

The small-business landscape is reflecting the female presence in an
emerging so-called maternal style of management. That supportive style
includes giving employees permission to have flextime, job sharing,
child-care subsidies, and maternity-leave benefits, and providing them
with an unspoken support system.

As a consequence, futurists John Naisbitt and Patricia Aburdene believe that under female influence managers will become "developers of personal growth, more facilitators than order-givers," and will introduce more "high touch" along with "high tech" influence. Thus, they forecast that there will ultimately "be more need for women employees, more competition for them, and more tolerance of different work patterns."[45]

NEW CULTURAL ATTITUDES IMPACT TELECOMMUTING

A change in generational values regarding work and leisure time, mobility, changing occupations, and family characteristics are impacting the work scene in general, and telecommuting in particular. For example, the family home is slowly but with deliberate speed becoming filled with electronic equipment and is but a computer step away from being the electronic cottage described in Alvin Toffler's *The Third Wave*. In his book, computers are used for home banking, to comparison shop, and to subscribe to information services, among other things.

More adults are currently remaining in the home. Many fathers are staying home to take part in rearing their children. Single parents are arranging to stay home and supervise their young children, and parents of teenagers are making more of an effort to be home during afternoon hours. This is part of the growing recognition that an adult presence can be vital in guiding teens during after-school hours.

People in the 18- to 24-year-old group have accepted technology and tinker with high-tech products, even maintaining their interest in the area when they have other career and financial responsibilities.[46] On the work front, their interest in technology is sometimes expressed by a show of loyalty to a technological system or career-oriented project rather than to the employer. For example, when a computer worker's particular task is not the corporation's priority and is allowed to lapse, the employee tends to become disenchanted with the company. "There's absolutely no doubt that corporate loyalty is lower today than it's ever been," according to one research specialist. It should be noted that the recessions and corporate mergers, layoffs, and other upheavals that have affected employees during the last years have contributed to this attitude.[47]

In some instances, employees are more interested in high technology than their managers are. It is not uncommon for a worker to buy a personal computer or other automated device in the hope of doing part of the office work more easily at home. This can have the unfortunate outcome of creating an attitude of resistance to the supervisor's or company's outmoded work style, leading the worker to become dependent on the arrangement and begin working outside of the job description. The situation can also discourage management from exploring the use of more efficient equipment in accomplishing office tasks. The ultimate danger is that computer work performed at home becomes inseparable from the employee's office work.[48]

A number of other evolving cultural influences are expected to affect American working styles and will certainly be reflected in the telecommuting picture. For example, companies are developing a do-it-yourself, "bypass" mentality by providing their own energy, communications, and transportation systems.[49] On an individual level, this is reflected in telecommuters becoming more entrepreneurial and bypassing agencies when they need work.

Within the general population, multiple, evolving career paths are replacing the traditional 20-year working arrangement, and entrepreneurship or group efforts are becoming increasingly important influences. In business, the supervisory role is changing, cross-coordination of managers is being emphasized, and the traditional management hierarchy is shrinking, effectively decreasing promotional opportunities for managers.[50]

Demographers project that within 10 years, 33 percent of the U.S. population (up from 20 percent today) will have a high-tech oriented lifestyle and seek more control over their work and their lives.[51] Department of Housing and Urban Development (HUD) economists expect that almost half of the U.S. population will eventually live in shared-space housing such as cooperatives or planned unit developments (PUDs). Theoretically, telecommuting is ideal for settings where home and technology can be shared.

The general population is also becoming increasingly comfortable with the idea of managing part or all of a work schedule in the home. Even Madame Mogul, offering star-guided advice to Pisceans in *The Courant* of Boulder, Colorado, warned that people involved in home-based businesses should do "some careful juggling so that it (work) won't interfere with the home life that is equally important to you."[52]

UNRESOLVED TELECOMMUTING ISSUES

The above major trends suggest a number of questions that relate to the future telecommuting picture. For example:

- Will America become the technology-minded, wired nation that futurists predict? Will interactive cable systems handle most banking and other financial transactions and enable users to order goods and services? Or, will direct broadcast satellites supersede cable connections and create a different scenario?
- Which segments of the work force will be telecommuting within the next 10 years as the remote-work picture evolves? Some experts believe that telecommuting will continue to evolve only slowly. Research for this book indicates that most people are inter-

ested in telecommuting, although there is no strong pressure to participate in remote-work programs full time.

- How well and how happily will telecommuters work over long periods of time?
- Will new managerial techniques and philosophy pervade?
- Is current equipment sufficiently friendly to encourage large-scale telecommuting programs? One telecommunications specialist reminds us that the telegraph was quite usable but the telephone proved to be "friendlier." Another believes that for a device to be accepted, it must be more useful or desirable or provide other rewards worth paying for.[53]
- How should a company advise telecommuters on their state and federal income tax deductions? The IRS is strict with employees who do business at home.[54] Companies need guidelines on resolving tax matters, insurance issues, and zoning laws.
- How will state and federal regulations eventually be written concerning the health, safety, and wages of telecommuters?
- To what extent will organizations be allowed to control the telecommuter working at home?

Throughout this book, we will suggest answers to these and other questions.

Conclusion

The changes taking place in technology, demography, and management structures are providing many of the driving forces behind telecommuting. At the same time, working conditions are becoming more compatible with employee needs. Among those needs is greater freedom within the employee's working conditions.

NOTES

1. Tammara H. Wolfgram, "The Right to Choose Where You Work," *Profiles,* May 1985, pp. 38, 60.

2. Tom Dworetzky, "Machines on the Go," *Discover,* July 1984, p. 77.

3. Margaret Eisen, "Ways to Profit from Businesses in the Home," *Computer Dealer,* March 1984, pp. 64–65.

4. Margaret Eisen, "Business Computing in the Home—How You Can Make It Happen," *Computer Dealer,* March 1984, pp. 62–68.

5. *Telecommuting Review: The Gordon Report,* vol. 2, no. 2, Gil Gordon Associates, Monmouth Junction, N.J., March 1, 1985, p. 6.

6. Patricia A. Bellew, "Technology" (column), *The Wall Street Journal,* May 10, 1985, p. 27.

7. Hal Hellman, "Home Sweet Office," *High Technology,* February 1985, pp. 64–66.

8. *Telecommuting Review: The Gordon Report*, vol. 2, no. 2, Gil Gordon Associates, Monmouth Junction, N.J., March 1, 1985, p. 8.

9. Matthew L. Wald, "The Smart Building," *The New York Times*, May 12, 1985, sec. 12, p. 1.

10. Walter Kleeman, Ph.D.; Francis Duffy, Ph.D.; Michele K. Williams, I.B.D.; Kirk P. Williams, I.B.D.; *Designing the Electronic Office: A Practical Guide*, to be published in 1986 by Van Nostrand Publishing Co.

11. *Telecommuting Review*, p. 14.

12. Susan Lindsay, "Computer Condominium Offered," *Systems Users*, May 1985, pp. 34–36.

13. Ibid.

14. Press Information, "Smarthome," CyberLINX Computer Products, Inc., Boulder, Colorado.

15. Eisen, "Business Computing," pp. 62–68.

16. Gregory Longhini, "Coping with High-Tech Headaches," *Planning*, March 1984, pp. 28–32.

17. Susan McAdams, "In Search of the 'Urban Telecommunications Infrastructure,' " *Scag Telecommunity*, May 1985, p. 2.

18. Kleeman et al., *Designing the Electronic Office*.

19. Penny Ward Moser, "Hooking Up in the Heartland," *Discover*, July 1984, pp. 81–84.

20. Kleeman et al., *Designing the Electronic Office*.

21. Peter F. Drucker, "Playing in the Information-Based 'Orchestra,' " *The Wall Street Journal*, June 4, 1985, p. 28.

22. Ibid.

23. Kevin Maney, "Companies Trim Fat, Fatten Profit," *USA Today*, July 3, 1985, Sect. B, pp. 1, 2.

24. John Naisbitt and Patricia Aburdene, "Good News: Here Come the New Megatrends!" *Ms.* magazine, July 1985, pp. 36, 37, 104.

25. Eisen, "Business Computing," pp. 62–68.

26. Philip J. Stone and Robert Luchetti, "Your Office Is Where You Are," *Harvard Business Review*, March–April 1985, pp. 102–17.

27. Ron Derven, "Plan Now for Emerging Markets," *High-Tech Marketing*, September 1984, p. 38.

28. Joann S. Lublin, "Running a Firm from Home Gives Women More Flexibility," *The Wall Street Journal*, December 31, 1984, p. 11.

29. Edward Wakin, "Jobs a la Carte," *Today's Office*, September 1984, pp. 43–47.

30. Lloyd Schwartz, "U.S. OKs 'Flexitime,' " *Management Information Systems Week*, May 29, 1985, p. 34.

31. "Flexible Hours," *New York Times*, July 2, 1985, p. 20.

32. Lad Kuzela, "More High-Tech Firms Turning to 'Temp' Employees," *Industry Week*, May 27, 1985, p. 33.

33. Dan Watkins, "Employee Leasing" (letter to the editor), *Business Journal*, May 27, 1985, p. 2.

34. Peggy Koenig, "Telecommuting: One Firm's Approach," *CommunicationsWeek*, April 8, 1985, Sec. C, pp. 1–2.

35. Naisbitt and Aburdene, "Good News," pp. 36, 37, 104.

36. "Want to Move?" (chart), *The Wall Street Journal*, October 8, 1985, p. 33.

37. *Land Use Digest* 18, no. 7 (July 15, 1985), p. 3.

38. The Yankee Group, *Yankee Ingenuity*™ newsletter, vol. 7, no. 3.

39. "Working Women" (chart), *The Wall Street Journal*, March 8, 1985, sec. 2, p. 1.

40. Derven, *"Plan Now,"* pp. 34–38.

41. *Telecommuting Review: The Gordon Report* (newsletter), vol. 1, no. 1 (October 31, 1984), pp. 1–10.

42. Joann S. Lublin, "Female Owners Try to Make Life Easier for Employees—Sometimes Too Easy," *The Wall Street Journal*, May 28, 1985, sec. 2-p.

43. Anne White Garland, "The Surprising Boom of Women Entrepreneurs," *Ms.*, July 1985, p. 94.

44. Bill Abrams, " 'Middle Generation' Growing More Concerned with Selves," *The Wall Street Journal*, January 21, 1982, p. 25.

45. Naisbitt and Aburdene, "Good News," pp. 36, 37, 104.

46. Derven, *"Plan Now,"* p. 37.

47. Thomas F. O'Boyle, "Loyalty Ebbs at Many Companies As Employees Grow Disillusioned," *The Wall Street Journal*, July 11, 1985, p. 29.

48. James T. Jarzab, " 'Home' Computing Perils," *InfoWorld*, Viewpoint (column), February 25, 1985, p. 8.

49. John Naisbitt and the Naisbitt Group, *The Year Ahead* (New York: AMACOM, 1985), p. 28.

50. *Sloan Management Review* 26, no. 2, pp. 45–49.

51. Neil Gluckin, "The Office Is Where the Workers Are," *Telecommunication Products + Technology*, June 1985, p. 60.

52. Madame Mogul (astrology column), *The Courant* (Boulder County, Colorado), June 12, 1985, p. 11.

53. Derven, *"Plan Now,"* pp. 34–38.

54. Herb Swartz, "Telecommuting's Expansion Poses Host of Legal Questions," *InformationWEEK*, April 15, 1985, p. 38.

3

Implementing
Telecommuting Programs

Telecommuting is in the position today that data processing was in 25 years ago. At that time, few business executives grasped the significance of the new technology, and possibly cost their companies a competitive marketing edge by losing time in adopting it. Of those executives who understood data processing potential, most did not know how to combine management issues with technology so they could achieve desired performance and cost efficiencies.

Today, organizations that lack bottom-line financial data on the benefits of telecommuting are also taking a wait-and-see posture and may be losing time and money in doing so. There is no question that "the [telecommuting] vehicle is coming into place. Now we have to figure out how to drive that vehicle," claims Rock Cary, information network manager at Apple Computer.[1]

The steps that are necessary to implement a productive remote-work program are discussed in this section of the book and cover the following issues:

- Upper-level management support and financial commitment for telecommuting objectives and for appropriate tasks (money is just not enough).
- Choice of consultants.
- Evaluation of current and proposed communications and information-networking systems.
- Promotion and expansion of a successfully integrated telecommuting program.

The following chapter discusses:

- Preparing for program implementation.
- Assessing and evaluating communications, information flow systems, and managerial techniques.
- Initiating pilot and training programs, with a focus on human resource and supervisory issues.
- Selection and training of suitable, motivated employees and skilled managers.
- Work agreements.
- Management/worker issues.

UPPER-MANAGEMENT SUPPORT

Because implementing telecommuting programs is costly and risks failure, upper-level management support is requisite. That support is gained by very clearly presenting the organization's remote-work options and the significant business case for using them over both the short- and long-term.

It is also desirable to have upper-level management involved in using a telecommuting system. Managers' participation promotes the acceptance of remote work among supervisors at all levels, as well as among other employees. While the groundwork for a remote-work program is being laid, those responsible should begin to build company interest in the topic and develop a foundation of prospective telecommuters.

Managerial discussion should first clarify the "real" versus the "perceived" benefits of a telecommuting program. An organization's real benefits are the measurable increases in an employee's production and the reductions in cost to the company when specific tasks are performed. Perceived benefits are usually described by such phrases as "It makes my job easier" or "I'm more comfortable working this way." Although real and perceived benefits are necessary for a successful telecommuting program, they should not be confused.

Upper-level discussion should then determine which tasks the company can best carry out, or which problems the company can solve by integrating a remote-work program, as well as specific job areas that are not appropriate for remote work. They involve activities that require employees to interact with clients, handle on-site materials, or interact face to face with people engaged in team projects.

The most persuasive argument for telecommuting is a hard-numbers, charted presentation of potential company savings, increased employee production, or time saved in doing jobs—whether those jobs are inputting data, writing advertising, designing computer graphics or programs, or simply transmitting raw information or messages via an electronic scanning system.

Whereas the knowledge worker's production, such as a project report or proposal, is usually calculated by tracking job objectives produced over time, a clerical worker's data or word processing input may be calculated by the length of time the employee's computer is connected to the company facility or by the number of strokes that are entered into the system each minute.

The point should be made during management discussions that a worker's increased production or well-organized informational output will create greater managerial accomplishment and shorten decision-making time. In addition, it should be noted that remote workers and supervisors who communicate via a teleconferencing system will experience less "down time," meaning those hours that are spent playing "telephone tag," traveling to and from the office for visits, and participating in face-to-face meetings.

However, companies must beware of choosing tasks for remote-work where key strokes or entering words and numbers are counted in determining production, while they ignore "soft," hard-to-evaluate tasks that require more complex work such as writing, financial analysis, and other knowledge work.

It should be emphasized during management discussions that because there are so many costly, critical factors involved in integrating telecommuting programs, they make a far better long-term organizational investment than a short-term, stopgap, measure. Indeed, without adequate "business case" planning and financial backing, a program may fail because management neglects to take sufficient time to:

- *Clarify* specific company/program objectives and methods of measuring them before deciding on remote-work tasks.
- Look *far enough* to find suitable consultants.
- *Completely* assess current and future bases for communications and information exchange on which telecommuting will depend.
- Establish *thoughtful (not arbitrary), well-defined* relationships between telecommuters and company.
- Develop *thorough* training programs for both managers and workers.
- Keep pilot programs *simple,* with *easily controlled* parameters.
- *Beware of self-selection* in choosing telecommuters and supervisors, always allowing employees the option of remaining in, or returning to, the office to work, *without prejudicing* their status.
- *Plan* for office space, services, and liaison for telecommuters who visit or work in the office part time.
- Provide telecommuters with a *fully supportive* program, meaning adequate supervision, feedback, regular communications and guidance, as well as a readiness to adjust the program itself.
- Anticipate and prepare for potential problems.

David Conrath of the Department of Management Sciences, University of Waterloo, Ontario, Canada, takes a somewhat different approach to adapting a computer-based support system like telecommuting. Conrath believes organizations should:[2]

- Start with a problem, not a solution.
- Think about people long before technology.
- Focus on support, not automation.
- Remember that support implies integration.
- The key to a system is integration into the organization.
- When considering side effects, do not think just about one step removed, think about two.
- Good management can survive without technology.
- The reverse is not true.
- We are basically social animals. Don't forget it!
- Effectiveness is essential, efficiency is not. Therefore, worry about doing the right thing (being effective) before doing things right (efficiently).

THE IMPORTANCE OF CONSULTANTS

The first step for an organization to take in implementing a telecommuting program is to set aside 5 to 7 percent of its proposed remote-work budget for consulting services. It is essential to find appropriate, experienced consultants, either from within the organization or recommended by other similar companies that have implemented telecommuting programs. The loss of money, time, and energy can be monumental for the company that is poorly advised on integrating a telecommuting program, particularly if it is forced to retrack and establish other operating modes.

Consultants can evaluate a company's current internal communications facilities and potential capabilities, then propose various hardware/software arrangements for the organization and present the business case for each of them. They should know vendor lead times and take responsibility for the contractual arrangements regarding the chosen system, as well as oversee its proper implementation. The consultant should also remain on hand to advise the organization during integration of the entire program.

Valuable advice can also be found within many organizations. For example, a data processing supervisor can inform management of telecommuting hardware/software that will interface with the company's computer facilities. Communications department personnel can explain how the company's telephone system can be used in a telecommuting program. Management information systems (MIS) people can make valuable contributions to discussion based on their experience with data processing, systems analysis, and computer programming. MIS employees, in fact, have proven to be "naturals" for remote work.

General office managers have valuable practical experience supervising people who use automated office systems. Companies tend to ignore people who work in personnel departments because they are often technology illiterate. This can be a mistake. A telecommuting program usually benefits from input from this source when decisions are made on remote job design, performance monitoring, worker job satisfaction, and benefits. Current general and specialized news coverage has alerted personnel people to changing pay scales, legal, psychological, and other issues regarding the remote worker.[3]

Marketing managers should also be drawn into discussion on proposed telecommuting programs. The factors that currently contribute to telecommuting growth affect long-range marketing survival.

DETERMINING APPROPRIATE TELECOMMUTING TASKS AND LOCATIONS

By carefully logging and analyzing work, then discussing it with other supervisors, a manager can determine whether a job, or only certain tasks within the job area, are suitable for telecommuting. It can be helpful to have the person who currently performs the job participate in its analysis. In this way, he or she becomes aware of the benefits of telecommuting arrangements, has the opportunity to offer valuable suggestions concerning the program, and can consider participating in it.

Management must remember that although telecommuting often answers the need for increased productivity, simply improving a supervisor's style of communicating sometimes works as well as installing technology-related devices. A key to determining remote job suitability involves knowing whether tasks are routine and often repeated, or done sporadically and in batches, and whether the work requires periods of total concentration, such as during the preparation of technical material, writing research reports, and handling certain other professional work.[4]

Managers must then decide whether the job or task lends itself better to a home work setting, a corporate satellite office environment, or elsewhere. In considering the location for remote work, management should be made aware that telecommuting can be handled in a variety of ways. For example, a typesetting firm in Plainfield, New Jersey, enlisted members of an Osborne Users' group to do data entry work when there was typesetting overflow.[5] The work was distributed to, and collected from, group leaders in each area. Each group of workers then picked up their material, completed the work, and returned it to the leaders' homes. No special benefits were given the group leaders.

An interview and a one-day training session in company standards prepared these "independent" (and intrastate) workers for the job. They earned $13 to $14 per hour, but received no other benefits. Company overhead was reduced, and staff members were freed to do more complex work.

ASSESSING CURRENT AND FUTURE TELECOMMUTING SYSTEMS

Before actively gearing up for a telecommuting program, it is essential that the people responsible for its integration carry through a detailed assessment of the organization's:

1. Current communications systems and their use.
2. Managerial methods.
3. Style of operating.

To fully understand these three areas, it is vital to hold a series of one-on-one fact-finding interviews at corporate headquarters with selected managers representing different areas of the organization, as well as with other key employees.

Where it is difficult to manage visits at headquarters, personal visits at field sites and telephone interviews may be in order. Interviewees should be solicited for specific, accurate information regarding workers' current productivity so that it can be analyzed and serve as a basis for both selecting the appropriate technology and justifying planned telecommuting costs. It can also serve as a base from which to compare each telecommuter's production and the remote-work program as a whole.

The interviewer should remember that the purpose of discussion is not to determine how people feel about a proposed telecommuting program but to gather an accurate picture of actual communicating, meeting, and working patterns. This information can usually be found in conversation that probes for an understanding of the individual's needs and the way in which he or she operates. During conversation, however, the interviewer can take the opportunity to describe the telecommuting process and its many benefits, although employees might not fully understand the picture and remain unenthused.

A thorough analysis of information gathered from interviews with employees should supply management with a full understanding of those structural aspects of the organization that relate to telecommuting. These include the following:

- Cross-functional and cross-hierarchical aspects of the organization's structure. Knowing these relationships can help managers use telecommuting (teleconferencing) techniques to expand their communications and information-sharing patterns.

- The organization's operating style. Style refers to the degree of formality in communicating and interacting among managers and other employees, and the amount and type of control that managers exert over employee work activities. In organizations where the style is very political, telecommuting success will depend on "who you know," rather than "what you're doing," according to authority Jack Nilles.[6]
- The context within which the organization operates. This refers to a company's industry, clientele and audience.
- Key work and commuting patterns among workers and managers. The attitudes of employees who must drive through central city congestion will differ sharply from those who can walk leisurely to work in five minutes.
- An analysis of employee work locations. In some situations, it is possible for a company to establish a decentralized office system or work center that serves telecommuters more effectively than a home setting.

Once upper-level management has considered the above general issues, it should determine whether the "corporate culture" is willing or interested in implementing a telecommuting program. Developing a business case will help gather management support.

If the company decides to implement a telecommuting program, it should remember that the user's needs are primary and that a thorough analysis of those requirements is vital in determining the equipment necessary for effective telecommuting. The results of this study will also help the organization determine the type of networking equipment required.

To complete such an assessment, it is also necessary to audit and analyze the organization's communications network by studying the following:

- Employees' preference for, and actual use of, various communications systems (voice, data, graphics, and text), how frequently they are used, the channels of transmission, and the costs of using them. For example, in analyzing employees' use of voice media, the organization must know whether people are holding simple telephone conversations or audio teleconferences. It must also know if the purpose of those communications is for data transmission from one computer terminal or device to another (as required in banking), for transmitting text from one device (word processor, computer, facsimile machine) and location to another, or for presenting videographics, as during meeting presentations.
- The organization's various communications networks (PBX, CENTREX, or local area networks), including their upkeep. These could be found in its internal offices, within an intrabuilding complex, at regional sites, subsidiaries, and branches. They could also

cover external networks, links with customers, a particular audience, organizations, or other business and professional connections.

- The organization's informational network. This network provides employees and management the means of accessing, storing, and transmitting information. The assessment should account for those waiting times required while information is moving along information network channels.
- Employees' (both workers' and managers') work and meeting matrix. This includes noting the type of meetings, the time they require (including planning), their relative importance, the accompanying text and graphics, as well as a description of the nature of that material and its purpose. The importance of flexibility in meeting schedules and consideration of "live" meetings versus nonreal time or retrieved store-forward (such as text or voice mail) communicating is equally important and should be noted.
- The organization's travel matrix. A study of the company's travel matrix should include employee destinations, the distances and scheduled times of travel, employees' attitudes towards travel, and possible telecommuting options. The opportunity for working with people at a distance and the cost of doing so should also be carefully examined. In figuring costs in this area, parking, telephone, baggage, and car rental charges must be considered, as well as interest lost on travel advances and on prepaid deposits.
- Technology involved in user communications. Technology issues encompass an organization's present office tools. These can range from such equipment as yellow (legal and other) pads and calculators to any of the smart office devices. In a large organization, these might include personal computers, teleterminals, and computerized facsimile devices, as well as such office support systems as electronic mail, voice storage, and micrographic equipment. This area is discussed extensively in Chapters 5 through 9.

ASSESSING TELECOMMUTING COSTS

In considering future costs in a telecommuting business case, the organization must assess the hardware and software necessary to accomplish both present and projected work loads as well as implementation of the technical side of the remote-work program. This assessment should also reflect the use of capital assets such as computers, current office space, and the control of current related operating expenses.

The city of Fort Collins, Colorado, found tools or services in the following three areas to be essential in supporting office workers:[7]

Telephone and access to city telephone features, such as:
   Long distance
   Conference calling
   Call forwarding
   Call waiting
   Electronic directories

Office automation that provides:
   Electronic mail
   Personal computing
   Word processing
   Electronic filing
   Calendaring/scheduling
   Mailing lists/distributing

Database/application access for:
   Financial reporting
   Financial transactions
   Other applications

Support for the telecommuting setting must be assessed in terms of providing comfortable heating, cooling, and ventilating systems, adequate lighting, humidity control, and the cost of ergonomic factors pertaining to furniture and work-area layout.

Likewise, the company must plan and account for a personnel training system that allows for an adequate number of telecommuting training hours and adjustment to program implementation so that proper work performance and evaluation are ensured.

Thus, by comparing present work system considerations and future telecommuting program alternatives, a financial analysis summary should enable management to pinpoint those departments and tasks with the greatest telecommuting potential. The areas to be covered would include:

Present work system considerations:
   Obsolescence
   Compatibility
   Service and support costs
   Staff training required
   Alternatives
   Security
   Management and organizational impact

Future telecommuting program alternatives:
   Capital expenses
   Implementation costs
   Recurring costs

Displaced costs
Sensitivity analysis (analyzing trade-offs)
Management and organizational impact

The Resource Guide at the end of the book provides a detailed outline of the assessment process.

Although workplace dangers are not part of a telecommuting financial assessment, the organization must be alert to avoiding them and the costs they may eventually create in any setting. Workplace contaminants for the sensitive or headache-prone remote worker can be a serious problem. Gases that contain formaldehyde are sometimes released from desktop laminates, new carpeting, plywood, and other materials, and giving some people headaches. Recirculated, warmed, or cooled air may also carry contaminants.

SECURITY OF INFORMATION DURING THE TELECOMMUTE

Clearly, the security of an organization's information begins by hiring trustworthy employees and keeping vital information confidential. It is also essential for organizations to establish trade-secret programs, provide sufficient internal safeguards, register equipment that is used, and have a written agreement as to the care and use of that equipment.[8]

On the technical side, a study of the communications tools in place is necessary to determine security considerations in integrating a telecommuting project. There are various technical means of controlling system access that guard against unwanted intrusion, as well as using fiber optic transmission systems and encryption (coding). Computers that have ID acknowledgement capability, for instance, allow company facilities to communicate with them. The use of a call-back system ensures that connections are made with authorized devices. Using only a simple dial-up capability to reach a company computer gives neither the worker nor the company sufficient protection.[9]

Although it may not be advisable for an organization to do payroll, accounts receivable, or handle stockholder records from a remote workstation, most corporations believe telecommuting poses no threat to their security once they have established the necessary precautions. IBM, for example, has placed more than 8,000 computers in employees' homes so they can work overtime, according to Electronic Service Unlimited.[10] It considers the only security danger to be in handling information when using "shared, shortened, or infrequently changed passwords."[11]

DEVELOPING ORGANIZATIONAL POLICIES
FOR A TELECOMMUTING PROGRAM

An organization that creates clear policies regarding its remote-work program lends that program legitimacy and supports those in charge. In order to promote and perpetuate successful telecommuting as well as give direction to department managers and "a sense of sanction" to departments that try creative approaches to job structuring, Fort Collins, Colorado, initiated the following policies:

1. Whenever possible, and to the extent practical, consider telecommuting when hiring, contracting, or otherwise securing services from new or existing personnel.
2. Encourage telecommuting to continue where it has previously been initiated. In order to promote greater citywide involvement, evaluations and recommendations for expansion of telecommuting should be communicated to the city manager.
3. Consider telecommunications as an alternative to written or verbal communication between city staff and vendors/project consultants.
4. Ensure that necessary funding and resources are provided for adequate equipment and facilities supporting employee telecommuters.
5. Distribute information and results from telecommuting projects to the public in order to promote and encourage further participation and broad involvement by the business and educational community.

EMPLOYER–EMPLOYEE REMOTE-WORK CONTRACTS

When integrating a telecommuting program, it is essential that the company and the remote workers clarify all job objectives and working conditions, as well as each person's responsibilities within the program. As noted in Chapter 1, a casual approach to program integration can be a major reason for its failure.

Any employer-employee agreement should include a statement describing working conditions, the equipment that will be used, and its care. It should also cover the company's expectations concerning work production, time parameters for completing tasks, and conditions for canceling the working arrangement after proper notification. It is vital that both management and remote workers closely track hours and work production.[12]

In establishing agreements with workers, management should note that a 1985 survey shows the following information is most important to employees:[13]

Employee benefit program	82%
Pay policies and procedures	78
Company's plans for the future	73
How to improve work performance	65
How the job fits in	61

However, an Equitable Life Assurance Society chart indicates that the following percentages of workers are dissatisfied with these aspects of their jobs:[14]

Salary	20%
Health plan	19
Benefits	17
Job	12

The manager should also consider the following issues and discuss them with employees:[15]

- Employee liability. At-home accidents occur in various situations, such as when the worker trips over wiring or a device falls on an employee's foot (also see page 64).
- Visits to check on safety measures.
- Confining work to one area.
- Checklist for the telecommuter that includes:

    1. Sufficient number of telephone lines.
    2. Well-grounded electrical connections.
    3. Appropriate equipment and furniture.
    4. Working area that affords privacy and safety.
    5. Proper and well-placed lighting.

A number of organizations rent telecommuting equipment to their employees. However, optimum efficiency and company loyalty increase when management supplies and pays for all necessary at-home office equipment such as the computer terminal, telephone lines, and good support services. Many employees appreciate being able to choose their own working hours.[16]

In discussing safety issues with the potential remote worker, a manager can use the opportunity to discuss any uneasiness the company may have about use of equipment in the environment and the position of the International Labor Organization (ILO) in regard to computer work. The ILO Advisory Committee on Salaried Employees and Professional Workers recommends that workplace conditions be regularly controlled to ensure employee comfort—particularly with regard to visual conditions. It recommends that employers provide "task variation"—that is, limit the time workers spend doing intensive work at terminals—and that workers be allowed regular breaks each hour when doing repetitious work. Although ILO studies indicate there is no conclusive proof that using computers creates any reproductive hazard, they ask that employers give consideration to the requests of pregnant women who want to transfer to other work.

Companies like Eastman Kodak and IBM have developed management guidelines intended to decrease VDT fatigue and glare, although these companies do not consider computers to be "health hazards," according to the executive director of the New York Committee for Occupational Safety and Health (NYCOSH).[17] Claims processors at Equitable Assurance Society have designed a contract they believe establishes monitoring limits.

Because computers have had the capacity to monitor employee productivity for some years and many organizations have a policy of monitoring employee work (whether employees know of this or not), the subject should be included in any preagreement worker/management discussion. It is widely accepted that there are two major benefits to monitoring employee production: It motivates employees to meet organizational standards or improve on them and helps the company rate a worker's performance without showing favoritism.

However, a growing number of labor experts, unions, and government officials are concerned that close monitoring creates undue stress for employees and makes them feel dehumanized. "It [monitoring] seeks to address workplace problems technologically when they perhaps should be addressed through better worker-manager relations," according to Harley Shaiken, a labor and technology specialist at Massachusetts Institute of Technology.[18]

As a result of possibly unfair monitoring practices, bizarre stories of computer sabotage have reached the courts. Surveys have found that computer workers with electronically monitored jobs were especially likely to report health problems such as chest pain, digestive problems, headaches, depression, and other complaints.[19] These could be considered a major source of stress. As a result of these diverse problems, a number of states are considering legislation that will restrict monitoring, and 20 unions have adopted official positions against it.

While many large organizations have currently established what they consider fair and equitable codes in relating to telecommuters, unions continue to be concerned with legislative standards or unionization that covers such issues as:

- Equal compensation for work done, regardless of location.
- Agreements or contracts covering equipment and its use, supplies, and similar costs.
- Fringe benefits and insurance.
- Federal and local laws.

On the management side of the telecommuting picture, a professor of business management maintains that, because there is often a machine between the user and boss or client and company, there can be difficulty in motivating personal accountability, and the system becomes vulnerable to operator control—particularly among disgruntled workers.[20]

Before employers and employees sign contracts that cover telecommuting conditions, organizations should clearly define those career paths that will remain available to workers once they have begun a remote-work program. Some employees will inadvisedly believe that accepting remote work puts them at an advantage for promotions, while others may fear that telecommuting will close the doors to such possibilities. Refusal to return to the office to work should also come up for discussion because it may affect personnel planning and Affirmative Action programs when promotion opportunities are involved.[21]

The media and trade publications have given considerable attention to the advantages and disadvantages of employees being paid on a per-piece basis. Indications are that most telecommuters do not object to working on that basis unless the employer fails to give periodic, fair pay increases.

Production quotas is another area for manager/worker contract discussion. In some companies, managers are concerned only about employee production. The hours worked are of little importance. In other companies, computers function only during given hours, such as 6:15 A.M. to 5 P.M., and work is paid for by the piece. Organizations that allow workers to choose their hours are known to grant those requests 70 to 80 percent of the time.[22]

Employee salary and fringe benefits can be negotiated when telecommuting agreements are drawn up. Companies that have an employee liaison system may give the telecommuter a voice when the organization considers salaries, benefits, and contractual agreements. As of the end of 1983, most companies were salarying telecommuters and allowing them the same fringe benefits as other office workers, according to Margrethe Olson, an associate professor at New York University's Graduate School of Business. Some specialists argue, however, that this is unfair because home workers are more productive and stand to gain when they are paid by the piece. Paying a bonus or higher rate for work produced is an alternative incentive managers can use to attract people to telecommuting.[23]

When discussing benefits and salaries, it is appropriate for the company to consider offering the worker alternative methods of compensation which may be more satisfactory for both. Job sharing and/or incentive plans that are organized on a contractual or goal-oriented basis would be one example. Management should not be surprised to hear unusual requests for employee support along the lines of child-care benefits or maid service. It has been rumored that some Japanese firms underwrite maid service for their telecommuters.

The company that fails to clarify its career advancement position can suffer near disasterous consequences, particularly when employees are ambitious and motivated.[24] For example, when the French planned to develop office work centers in the suburbs around Paris, employees of a given company were expected to work in them as groups. The company firmly believed that their employees would opt to save time and work close to home, and that their need to socialize would be satisfied by this arrangement. They misjudged, however. Employees who were given the option chose to continue traveling extra hours to the city's center so that their presence in the company's offices was recognized.[25]

For the potential telecommuter or entrepreneur, there is no substitute for the financial advice of a tax attorney to determine the benefits or liabilities of one's position. Many organizations now provide their telecommuters with this service. Workers who live at a distance from the area in which a company is headquartered may be covered by different taxation and zoning arrangements. It therefore behooves management and employee's alike to give careful consideration to possible tax problems that might arise as a consequence.

Internal Revenue Code terms are also a primary issue. The picture encompasses such matters as home office deductions, the depreciation of business purchases, tax credits that are allowable, and a bonus for a first-year write-off of equipment. *Standard Federal Tax Reports-Number 2* (January 9, 1985) indicates that the courts may be more supportive in the future of people who must work at home to have the privacy, safety, and additional space their tasks require.

There remain the problems of a company's right to control the employee in his own home and the stress in the employee-management relationship over problems of company control, taxation, benefits, and independent contractor status.[26]

Responsibility for abiding by local zoning restrictions generally belongs to employees. If the remote worker's home is considered a secondary work location, there should be no legal problem. The employee is usually responsible for converting a home for telecommuting, although telecommuting and related costs remain the employer's.

THE REMOTE-WORK SETTING

Because the design and facilities in any work environment influence employees and their productivity and can project an atmosphere that conflicts with remote-work goals, the manager and telecommuter must assess the home or work center setting before initiating a telecommuting program. The ideal workplace should be arranged and evaluated with the following objectives clearly in mind:

- To securely carry through the particular telecommuting project using the most technologically advanced equipment. (Chapters 6 through 9 discuss telecommuting technologies.)
- To provide a suitable work setting. This means using ergonomic furniture that is functional, adaptable, and aesthetic in appearance. It has been said that good furniture can effectively increase

an employee's production by an average 12 percent, but there must be suitable, adequate storage space as well.[27] A State Farm Mutual Automobile Companies (an insurance company) study reflected a 15 percent increase in production where ergonomic relationships in the work setting provide greater efficiency.[28] The working area should be spacious but not overly large. Too large a working space distracts some people. It is important that there be ample storage and counter space, particularly when the worker must copy material from large ledgers or several resources. (The work setting is discussed at length in Chapter 5.)

- To create a private and pleasant work atmosphere, a place that the family must respect as being off-limits during working hours. This must be assured if the worker is to concentrate on the tasks at hand. In addition, futurist John Naisbitt reminds managers that high-technology activity requires "high-touch" balance, or the technology will be rejected.[29] Therefore, the work setting should blend equipment with furniture that has a homey character.

The work center setting should include these additional features:

- Windows and artwork, if possible. A restful, natural setting as a counterpoint to the tedious work effort can provide a more productive environment.
- Acoustical insulation for loud printing devices.
- Computer room if space is short.
- Grouping of workers by task. Data entry requires less space than word processing.

An organization does not always realize the amount of equipment and space that is necessary for an effective work-at-home project. A recent study found that an oil company analyst required: a personal computer, hard disk drive, printer, monitor, modem, telecopier, telephone, telephone answering machine, white board, desk, chair, typewriter, filing cabinet, and lighting fixtures. When one or another such item is required, an organization can turn to the specialized suppliers who provide work-site equipment ranging from the most complicated of teleconferencing devices to the last stick of office furniture—and the personnel to advise clients on using it.

Home office and remote-site design is of increasing general interest. The National Endowment for the Arts has given the University of Southern California a grant to study emerging trends in new-home design and modifications that are being made to existing houses to accommodate the telecommuter.[30]

EXPANDING THE SYSTEM

Once a telecommuting program is well under way and the organization decides to expand the system, it will be necessary to:

- Closely monitor the use of the system, checking on trends in its use and workers' comments.
- Assess the state of development of telecommuting against company expectations in all areas (communications, production, employee response, and satisfaction).
- Develop plans for expansion, modifying telecommuting settings according to the actual requirements of remote-work tasks and for features that users have requested.
- Look for new users and applications, being sure to ask current managers and telecommuters for suggestions.
- Continue advertising the system by writing it up for internal company organs, company bulletin boards, the local and trade press.
- Check on the current state of telecommuting technology to learn if there is a possibility of upgrading any part of the system.

Conclusion

Implementation of telecommuting is only as good as management's commitment to the project on the one hand and the motivation and capability of the employees on the other. Telecommuting has been very successfully managed in some organizations as well as a dismal failure in others. Therefore those companies seriously interested in integrating a remote program must carefully develop plans for each step along the way, correct their mistakes quickly when they stumble, and when the program works, tell the world!

NOTES

1. Margo Downing-Faircloth, "Would Working at Home Be Wise?" *Personal Computing,* May 1982, p. 42.

2. David W. Conrath, "Office Automation: The Organization and Integration," *OAC '85 Conference Digest* (Atlanta, Ga.: afips PRESS) pp. 310–11.

3. *Telecommuting Review: The Gordon Report* vol. 2, no. 2, Gil Gordon Associates, Monmouth Junction, N.J., March 1, 1985, p. 4.

4. Dr. Evan Peelle, "How to Make Telecommuting Work," *Personal Computing,* May 1982, pp. 38–40.

5. *Telecommuting Review,* p. 2.

6. Michael Antonoff, "The Push for Telecommuting," *Personal Computing,* July 1985, p. 91.

7. Peter K. Dallow, *Telecommuting in Fort Collins: A Case Study,* October 15, 1985.

8. Gerardine DeSanctis, "A Telecommuting Primer," October 1983, pp. 214–20.

9. Ibid.

10. *The Wall Street Journal,* February 5, 1985, p. 1.

11. *Telecommuting Review: The Gordon Report* (newsletter), vol. 1, no. 1 (October 31, 1984), p. 7.

12. DeSanctis, "Telecommuting Primer," pp. 214–20.

13. "Tell Me More," (Source: The Hay Group for Management Database), *The Wall Street Journal,* August 9, 1985, p. 19.

14. "Low Morale," Equitable Life Assurance Society chart, *The Wall Street Journal,* September 9, 1985, p. 23.

15. Jeanne Woldenberg, "Telecommuting: No Workplace Like Home," *Words,* June–July 1984, pp. 24–27.

16. Marilyn Webb, "Life in the Electronic Cottage," *Working Woman,* December 1983, pp. 106–109.

17. Emily Leinfuss, "Labor Group Hits VDT Hazards," *Management Information Systems Week,* May 29, 1985, p. 34.

18. Michael W. Miller, "Computers Keep Eye on Workers and See if They Perform Well," *The Wall Street Journal,* June 3, 1985, pp. 1, 15.

19. Ibid.

20. Wang Laboratories, Inc., *On Human Factors* (Lowell, Mass. 1983), p. 8.

21. DeSanctis, "Telecommuting Primer," pp. 214–20.

22. Eric R. Chabrow, "Telecommuting: Managing the Remote Workplace," *InformationWEEK,* April 15, 1985, pp. 27–35.

23. Webb, "Life in the Electronic Cottage," pp. 106–109.

24. Wang Laboratories, *On Human Factors,* p. 5.

25. Conrath, "Office Automation," pp. 310–11.

26. Swartz, "Telecommuting's Expansion," p. 38.

27. *Telecommuting Review,* October 31, 1984, p. 3.

28. Ivan Cutler, "Designing the Office around the Computer," *Personal Computing,* November 1984, pp. 66–70.

29. John Naisbitt, *Megatrends; Ten New Directions Transforming Our Lives* (New York: Warner Books, 1982).

30. *Telecommuting Review,* October 31, 1984, p. 8.

4

Understanding the Human Factors of Telecommuting

INTRODUCING THE SYSTEM

The organization should promote an early awareness of its telecommuting program among employees by giving the program a name (perhaps a humorous one), advertising it in company newsletters and local magazines, and describing it on posters and bulletin boards. A telephone number for additional information and support should be included in promotional material for interested employees. A telephone answering machine for extended business hour coverage should also be provided, as well as a number to call for more personal "help" once the program is initiated.

The organization might inaugurate the telecommuting program by holding a "grand opening" and monthly receptions for staff members who are being initiated into remote work training or pilot programs. Where it is appropriate, people who were interviewed when the organization's communications systems were analyzed can be called back and asked to participate in the program.

Promotional help can be solicited from within the corporate communications, graphics, and marketing departments. These departments can assist in developing user guides that explain telecommuting systems and include information on company policies that pertain to those systems.

TELECOMMUTING PILOT PROGRAMS

It is the pilot program that gives an organization the chance, without undue risk, to:

1. Assess the economic reality of telecommuting.

2. Determine the level of productivity that can be expected from participants.
3. Review manager-employee interaction.

Specifically, a pilot program feasibility study enables management to determine:

- The effectiveness and efficiency of the technology used.
- Task parameters that guide program integration.
- People who work well in a home or work-center setting.
- Supervisors capable of managing a telecommuting program.
- The best procedures to use in establishing management-telecommuter relations and agreements.
- Successful managing and communications techniques to use in setting work or professional goals.
- All legal aspects of the situation.
- The effect of remote work on participants. This can be determined by evaluating whether the project has developed as the manager and telecommuter expected, whether working conditions and agreements are adequate for the employee's circumstances, and whether project goals can be met to management's satisfaction, or modifications will be required.

Pilot project objectives must be simple and clearly defined so that they can be easily evaluated in terms of office production and turnaround time. The results should enable the company to assess which hardware and software are essential as well as create a personnel management system that ensures productivity and allows performance evaluation. This is important because employee productivity during a remote-work program tends to increase in the short-term, but often requires new methods to sustain it over the longer haul.

A pilot program enables management to gain a firsthand understanding of the at-home worker experience and the advantages and problems created by this type of working environment for particular tasks and for the people involved. It provides the company with an opportunity to learn how to operate so that employees' corporate visibility is not affected. Many present-day field office workers have experienced being out of sight and therefore out of management's mind when it is time for promotion. As a result, they have had to consider returning to the office to maintain their visibility.

It is unfortunate that company departments neglect to run pilot projects when the telecommuting concept is widely accepted within the organization. They fail to realize the range of benefits that pilot projects provide, that they are not necessarily expensive, and that they can be easily executed on a small scale with little risk.

MANAGEMENT/WORKER SUPPORT AND TRAINING

There are a number of avenues that an organization can take to support managers and workers during implementation of a telecommuting pilot program. These include developing guides, providing the worker with several of the very good books written on the subject, and holding sessions on such topics as:

- Managing the technology or equipment to be employed.
- The physical design and conditions of the telecommuting setting.
- Time management for the projects undertaken.
- Developing one's self-management skills.
- Handling the social home environment.
- Psychological problems that typically arise during remote-work programs.
- Company policy issues and benefits.
- Legal aspects of remote work; zoning and local, occupational, and other taxes.

It should be noted that while employees may complain that their companies neglect to involve them in the development of telecommuting programs, studies have shown that excessive employee involvement in implementing office technology can produce weaknesses in the projects that are undertaken. Both management and worker must remember that it takes job planning and design as well as many weeks of work to develop a smooth-running program and working relationship between a supervisor and remote worker.

Training should be geared to the users' capabilities and designed to hold their attention. Brief, colorful video presentations, lively and to the point, are known to have impact. Repeat, hands-on sessions are often necessary before a trainee appreciates the benefits of the new system. Hand holding and close guidance during training sessions also may be essential for some new users. But, as in all orientation situations, trainers must listen attentively for the trainees' feedback.

A training program can help the worker in transition from an automated office environment to the home setting. Even when people have been doing a good job, many fear "change, fear the unknown; technology itself, that you'll make mistakes and look foolish or break the machine; (suffer) status loss, because office automation changes the rules; economic loss" that demeans them or eliminates their jobs, according to a consulting firm president whose company specializes in office productivity.[1] See the Resource Guide at the end of this book for good books, magazines, and newsletters that can help the telecommuter adjust to new circumstances.

While extensive training for remote work is costly, telecommuting specialist Jack Nilles contends that the cost is well offset by training benefits. In addition to the above reasons, he believes good management and an effective training program are essential for overcoming "adversarial relationship(s)" that tend to develop between workers and managers. For example, secretaries often want to take on more administrative tasks, but managers doubt that they have the ability or time available to do so. Or secretaries may want better communication while managers want less interruption. And whereas managers may trust automation for efficiency, workers may believe it develops through teamwork.[2] Electronic Services Unlimited is one of the companies that provides "tailored training services to companies interested in telecommuting programs."[3]

An orientation program that coaches telecommuters on life style issues will benefit the worker directly and the company indirectly. A major issue is the employee's home relationships. The family or household must show respect for the worker's privacy, space, and need for quiet. If it does not, the worker cannot concentrate and job tasks are not completed.

The problem arises because few families associate home with structured work schedules, total privacy, and concentrated work effort. Rather, they regard home as the place one goes *after* work to relax, and the place for family togetherness. In addition, people often believe the telecommuter has all the hours of the day or evening available to work and therefore will not mind being interrupted. An occasional telecommuting workaholic may, however, demand too much family cooperation and impose a rigorous, unbalanced schedule on the household.

Telecommuters, like other workers, must balance their hours of work and relaxation. People who overextend themselves generally burn out fast. Or when the self-imposed work regimen and pleasure of seclusion becomes a trial of loneliness, the work setting produces stress, causing the worker to have such symptoms as lower-back pain, headaches, indigestion, and insomnia.[4]

Some telecommuters are unable to draw the line between house work and remote work and feel guilty if they neglect their home or family chores during telecommuting hours. In addition, marital difficulties tend to emerge where couples have unstable relationships and have not previously spent long daytime hours together over extended periods of time.

In regard to the physical side of a highly automated work setting, a management studies professor cautions people to remember that "as we get into more abstract work, knowledge work, we lose a great deal of the sensation of our work." This creates the human need for more physical activity, like jogging.[5]

There is a myth that telecommuting allows the home worker to keep an eye on a baby or young child while engaged at the computer. It is generally acknowledged that such double duty is usually unsuccessful and baby sitters are necessary.

Although organizations may provide managers and remote workers with good training during implementation of telecommuting programs, management must often continue, over time, to caution workers on keeping a balanced lifestyle, keeping control of the home-work situation, and guarding against the following typical problems:

- Retreating into work exclusively and failing to socialize.[6] This often lowers morale.
- Expecting too much of themselves while being their own boss and burning out, or creating a chronic stress pattern that leads to poor health. Workers under stress often find their health deteriorates or becomes a consuming issue for them. Perfectionists who tend to give themselves negative feedback and achievers appear prone to such a pattern.[7]
- Shifting of normal time schedules can create health problems for the telecommuter and scheduling problems for other family members.

Other potential problem areas that may come up for repeat discussion include:

- Weight gain—the refrigerator is always close by.
- Fear of "stepping off" the corporate ladder and becoming less promotable or losing career opportunities.
- Being exploited.
- Being unable to exchange ideas face to face.
- Developing a smoking habit.
- Requiring more corporate "visibility" and recognition. Unfortunately, planners often fail to account for a remote worker's fear of being out of sight and thus out of mind.

CHOOSING TELECOMMUTERS

Because employee attitudes toward office and knowledge work vary widely, the organization that establishes a remote-work program must take great caution in choosing personality types that are appropriate for telecommuting. At one extreme there are people who love their work and immerse themselves in job tasks, forcing family life to revolve around the job. At the other extreme are office workers who see their jobs as part of the office scene, who cannot operate without dressing up for the work day, who look forward to observing office politics, partaking in company gossip or information exchange, and lunching with friends.

The worker who adapts best to at-home or work center telecommuting (noted in Chapter 1) is probably already working successfully in an occupation that requires independent, organized operating methods and extended periods of concentrated effort. Such people are usually more achievement oriented, plan well, are persistent in their endeavors, and self-directed.[8] In addition, they balance their time and can say no to intrusion.[9] They are often found among an organization's highly motivated, disciplined, reliable staff.

Many successful telecommuters are openly disdainful of group activities and committees, but display a sense of purpose, courage, responsibility, and little fear of failure in doing their tasks. In addition, they generally do the undesirable tasks first and find their greatest reward in completing their work well.[10]

Individuals who are true self managers—regardless of organizational level—will have the following characteristics:[11]

1. A desire to master new skills and continue learning.
2. The ability to assess their own performance or feedback.
3. The ability to set priorities and meet deadlines.

The professionally trained individual who must remain at home to take care of a handicapped person or other family member often fits this niche.

Before choosing remote workers, a company should investigate available personality inventories or questionnaires that help employees identify those personality traits that will support or hinder them in a telecommuting situation. Questionnaires can also help a company determine whether employees enjoy, and are motivated to do, the type of work that will be tackled at home, whether they have a balanced attitude toward remote work, and if they expect to be successful in doing it.

For the achievement-oriented person, companies should remember that it is "recognition for achievement, interesting work, increased responsibility, and personal growth" that take precedence in work, according to Professor Frederick Herzberg, writer on management theory.[12] When testing indicates that a worker is both comfortable with automated office equipment and a good candidate for remote work, the employer certainly has reason to be pleased.

When a company allows too much self-selection in choosing people for remote work, those who have not worked alone for extended periods may underestimate the feelings of isolation a telecommuter can experience. For example, those who work in groups or in large, busy offices often believe they are excellent candidates for telecommuting. They imagine the home setting as a place to work privately, without interruption, and to get their work done—until they have that time and are without the presence of co-workers to chat and exchange ideas with and are without the secretarial and other services on which an office depends.

People who are not self-sufficient in their approach to work will miss the constant guidance available in an office setting and fear their inability to manage themselves.[13] This can lead to procrastination in tackling remote tasks or to an overeagerness or workaholic attitude developed in an effort to prove oneself. Both problems can lead to stress and burnout.

CHOOSING THE WRONG TELECOMMUTERS

In choosing employees for a telecommuting program, there are a number of traps into which managers can fall. The supervisor can easily underestimate an employee's need for social interaction by failing to identify the socially active or dependent personality type and placing him or her in the isolation of a home setting. Such people are generally "outer-directed" individuals who depend on supervision for guidance and, hoping to please, may accept the manager's suggestion to work remotely. Fortunately, very sociable people often realize that they do better work and are more content in a lively social setting and will self-select out of a remote work program.

Similarly, it is easy for a manager to enlist enthusiastic new employees in a telecommuting program before they have a sufficient knowledge of the company and before their working habits and personality characteristics are clearly apparent.

Managers must beware that employees in certain office environments may be driven to request telecommuting in an effort to save their jobs, although they are not appropriate telecommuters and may realize it. This can happen when office personnel do not get along well and conflicts develop, when department "favorites" receive more attention than other workers, and where managers create disruption. As one telecommuter who requested anonymity said, "My boss keeps me from doing my job. I like the job and need it, but don't know what to do."

People respond to the remote-work situation in a variety of ways, some of them quite unexpected. There will always be a percentage of people who appear to be good potential telecommuters, but who will grow to feel isolated at home, develop problems, and ask to return to the office to work at least part time. Others will continue to feel close to their co-workers because electronic messaging keeps them in touch. Certain individuals will thrive on making their own decisions in the remote-work setting, others will find it stressful to do so.[14] Few telecommuters realize what it means to work without the office support system or to exchange face-to-face meetings for electronic messaging until they must do so.

MANAGERIAL LEADERSHIP ISSUES

Skillful management is the key to successful telecommuting programs. Supervisors must have sufficient background and know-how to deal with typical remote-work issues and avoid the traps that lay in wait for such a program. The manager who has had personal experience telecommuting should have a clear understanding of remote-work issues and, hopefully, the vocabulary with which to discuss problems that typically arise. One research director suggests that supervisors work for a period of time with their subordinates in the remote-work setting so they can relate to the situation and see how task details are executed.[15]

Certainly, supervising telecommuters requires the manager to change his or her leadership style and develop new skills. The first skill is guiding the worker who is absent. This is done by clearly defining the goals expected of the worker and the time limits allowed for particular company projects, by establishing communications channels so that both parties are easily reached, and by practicing good communications skills. Upgrading communications skills also helps the manager better relate to in-office workers as well.[16]

It is vital that people who manage a telecommuting program establish trusting relationships with their subordinates and provide them with sufficient feedback to know just where they stand. "I believe the future [in telecommuting] for all of us is in a free, trusting relationship," and that given time, trials, and failures, work at home can be successful, according to a director of training operations for Mountain Bell.[17] Part of the difficulty in doing this comes in establishing different performance measurements for telecommuters than for office personnel.[18] Regular but not excessive communication contributes to this trust, according to writer William Atkinson.[19]

Managers who operate as facilitators rather than order givers are generally more successful in handling remote workers. They show telecommuters how to deal with such important issues as:

- Organizing blocks of work time.
- Separating their personal from their work life.
- Making the most of time spent in the office, including taking advantage of social activities so they do not feel isolated.
- Accepting the importance of the remote-work period as a trial situation.

Being clearly understood is an issue that can haunt managers who are beginning pilot telecommuting programs.[20] They must take pains to communicate clearly when delineating instructions and then listen closely for the trainee's or remote worker's response. To guarantee both the managers' and workers' undivided attention in discussing delicate or complex instructional or other matters, they are best discussed face-to-face in the office or home setting rather than on the telephone or by electronic mail.

Listening is an undervalued side of communicating. Poor listening skills can create a serious communications problem for both managers and workers. According to writer Arthur Miller, poor listeners will tend to focus on the facts of discussion rather than the significant train of thought. They also will tend to prematurely evaluate the situation under discussion, cut in with questions, jump ahead of the speaker, and allow themselves to be distracted.[21]

Unfortunately, many upper- and middle-level managers find it difficult to supervise subordinates who are "out there" when they are used to, and prefer, face-to-face interaction and the give-and-take of decision making with peers—along with the approval and power these situations confer. This type of manager may fear that workers are out of control and losing interest in their tasks, watching television or doing housework rather than tackling the work for which they are paid. It is for this reason that a foundation of trust must be created between the supervisor and remote worker. There are also managers who fear that they will not be considered true managers because their staffs are working remotely and responsibly without *visible* means of control.

When the supervisor of a remote-work program is not readily available to subordinates, he or she both creates a deep sense of frustration and causes the worker to lose working hours. Supervisors must determine how to keep communications channels open, parcel their time among in-house and remote workers, and allocate time and office space for telecommuters when they work in the office.

The manager who imparts to the employee a sense of belonging to the organizational team and being in control of the work effort can improve employer-employee relations as well as the quality of employee production, according to authors Peters and Austin. "People who are part of the team, who feel as though they 'own' the company and 'own' their job, regularly perform a thousand percent better than the rest," and perceive that they have a modicum of control over their destiny.[22] Without a cohesive spirit, many remote workers are less motivated and feel little common purpose with co-workers back in the office. This attitude can affect both the quality and quantity of their work.

There are a number of ways that management can firm these bonds. Among them are: spending extra time with employees, listening closely for feedback on the work arrangement, asking for suggestions for change or better products and processes, and responding to complaints. The company can also send the remote worker publications of interest and invitations to company events or meetings. In addition, it is important that supervisors encourage employees to interact with the regular office personnel. The company that successfully integrates telecommuting requires a "social glue" that bonds people into a certain commitment of spirit while they work hard for a respected boss, share the work load with co-workers, and are recognized for doing good work.[23]

A telecommuters' club or group within the company can provide the remote worker with a welcome support system. Such an organization gives people confidence to succeed by providing quick answers to the simple questions that the new telecommuter typically asks. It also assures the individual that the automated system really can be easily managed, according to Carl Schlaphoff, a systems designer.[24]

When telecommuters communicate via electronic mail, the organization experiences valuable side effects, according to communications specialist Dr. Charles Steinfield.[25] He found, for example, that although electronic mail systems are used far more often for circulating information about interesting things or events than for private or confidential (company) information, employees who use them to contact company friends will feel closer to and more integrated into the organization and are less likely to experience "drastic shifts in working patterns." In addition, Steinfield's study found that employees who felt a need to communicate would find a way to subvert rigid productivity-oriented frameworks imposed on their telecommunications systems and build a "play or pleasure component" into them.

The advantages of office visits and personal contacts are quite apparent. They allow employees to keep touch with the office environment and visit with friends. "Being there" also gives them visibility and the opportunity to review their working arrangements, straighten out any job difficulties that arise, and pick up their mail.[26] Contact with the office also allows employees to socialize with friends and keep open their lines of communication. Studies show this is important because most employees take a lively interest in company gossip and expect to hear it via:[27]

- The grapevine—66%.
- A supervisor—47%.
- Company publications—42%.
- Memos—40%.
- Bulletin boards—38%.
- Management meetings—29%.

Office visiting rules for telecommuters naturally depend on the needs of both the organization and the remote workers and thus vary widely from one company to another. At one extreme, there are programs that limit employee telecommuting to one or two days per week and confine the work to strictly clerical-type tasks. The employee is in the office the rest of the week. At the other extreme, organizations such as Blue Cross/Blue Shield of Washington, D.C., have full-time telecommuters who come into the office only once a year to review procedures.

J. C. Penney Company, Inc., holds procedural briefings for its home workers once a month during the peak pre-Christmas season and once every two months during the rest of the year.[28]

It is often productive for the remote-work program managers to meet with their workers in the home setting. At that time, personal and office problems can be discussed privately, without office distraction or interruptions, and the supervisor can focus more personal attention on employees and project discussion.

Successful automation can create problems too, according to a Kaiser-Permanente office supervisor. "If it (a program) is not well supervised, there can be a tendency to take advantage." To avoid the problem, she suggests that management expect people to produce more in the automated environment but pay them well for their work and challenge them to be creative in using the technology.[29]

It is important that an organization balance remote and office work. The mail department of New York Telephone Company has balanced its program by having personnel divide their time between work at home that requires concentration—such as writing reports and memos, analyzing data, preparing legal briefs, reading job applications and similar applications—and interactive work that must be done in the office. The latter includes holding meetings.[30]

Honeywell, Inc., Minneapolis, has allowed some employees considerable leverage in choosing the hours during which they telecommute. For example, a public relations manager who lived about 25 miles from headquarters was allowed to spend most of her workweek at home caring for her new baby. She found it convenient to schedule her work between 4 A.M. and 9 A.M., using her computer access to telephone lines to send her work into the office for review. She also sent electronic messages to people who worked for her in Boston, Minneapolis, and Phoenix.[31]

In facilitating a telecommuting program, it is vital that people who supervise remote workers think innovatively and encourage their subordinates to do so. (Creative management concepts and evolving management positions are discussed in Chapters 2 and 3.) This book suggests that management specialist Peter Drucker's approach to innovation in business can be used to enhance a telecommuting program.[32] Drucker believes that organizations should hold candid junior-senior management discussions about the company and its field or industry. During those discussions, upper management can clearly indicate that the organization is ready to discontinue nonproductive processes, products, and activities and is receptive to new ways of doing things. Drucker also believes that management should use this opportunity to emphasize that some measure of creativity is expected as part of the employee's job.

When people are comfortable with new technology and feel encouraged to use it freely, they often apply it in surprising ways. One study indicated that people had used electronic mail for more than 30 purposes, a large proportion of them unanticipated by the company. When application programmers used an electronic mail system to play games, for example, the participants found the solution to a problem they encountered in developing product-oriented software. In another instance, employees placed questions on a public bulletin board as a way of reaching an individual whose calls were being screened.[33]

Organizations dare not forget that exchanging information, brainstorming, and cooperative problem solving not only improve the quality of work produced but motivate and give participants a feeling of common purpose with their fellow workers. Management can support this objective by creating self-managed work teams and problem-solving groups as well as by holding information sessions where the company's progress, future directions, and concerns are shared. In the social area, activities like team sports, picnics, and volunteer, civic-oriented projects carried out during leisure hours help build desirable social bonds among employees.

MANAGERIAL PROBLEMS

It is common for the manager to misjudge certain aspects of a telecommuting task or the employee's attitude. For example, a supervisor may overestimate the amount of time that a worker must be on site to handle office matters, meet personally with clients, or interact with fellow workers. Likewise, the supervisor may tend to underestimate the worker's need to socialize, according to Dr. Evan Peelle, organizational psychologist and management consultant.[34]

Problems inevitably arise when the office manager presents telecommuting to prospective remote workers as key to a promotion or a raise and neglects to say that one's job status is not affected if the opportunity is rejected. As a result, there have been instances when people who chose to remain in the office believed they lacked status, that the company imposed on them by interrupting their work frequently for phone answering, that they were required to do more rush jobs, and were forced to perform under more pressure than their telecommuting counterparts. Office workers may, however, simply envy the telecommuter's freedom from supervision.

Some managers are prone to forget the telecommuter who is doing a creditable job out of sight and away from the office, especially when there is someone readily available for the office job slot just above. Supervisors who allow this to happen will most certainly create personnel and possibly legal problems for themselves as well as for their companies.

"Sunlighting" is a worry for certain managers. This refers to holding two telecommuting jobs simultaneously or telecommuting along with doing other work, such as telephone answering or providing child day care. Managers who closely control their subordinates are generally insecure when they hear of such arrangements, even when the remote worker is fulfilling company obligations. One writer suggests that some managers fear the sunlighter may make more money than they do.[35]

"Flaming" is a well-known problem among people who use electronic mail systems. Flaming refers to the anger that terse or ambiguous electronic messages can create, often without the sender realizing it. The problem can be avoided to a degree, however, if management thoroughly discusses the issue and designs a code of etiquette to use in transmitting messages. Research done at the Carnegie-Mellon University suggests a number of explanations for flaming that arises during teleconferencing communications:[36]

- It takes people longer to reach consensus electronically than "in person."
- "Computer-mediated communication is depersonalized," providing no visual feedback during electronic conversations. Thus, participants can't see that they are understood.
- The moderator or leader has less influence.
- There is little or no shared etiquette or format norms for participants.
- On electronic teleconferencing systems, where people can communicate many to many (teleconferencing systems), one to many (broadcasting), and one to one (electronic mail), there can be significant changes in the character and tone of "conversation."

Just as organizations must establish workplace rules for headquarters office and home settings, they must do so at remote-work centers or chance creating conflict among employees. Surveys strongly suggest that smoking is one potential trouble area that can lead to problems of worker morale. Where, for example, some employees may insist on smoking during working hours, their nonsmoking co-workers will just as surely object.[37]

THE WELL-INTEGRATED TELECOMMUTING PROGRAM

As mentioned earlier, hundreds of telecommuting programs are running successfully today. Metropolitan Life Insurance Company management is among those credited with integrating a logical, structured approach to remote work.[38] The company began its remote-work program by hiring several trained handicapped workers who traveled to work one day of the week "to receive assignments, pick up mail, and establish needed face-to-face contact with managers and co-workers." They telecommuted during the other four workdays. While implementing the program, the company was able to resolve the following issues:

- An appropriate pay scale.
- Insurance coverage.
- Supervisor's delegation of time among employees working in-house and telecommuting.
- Allocation of office space for days the telecommuter is working in the office.
- Communication. The project manager first used the telephone and then shifted to communicating by computer.

Conclusion

As we mentioned at the end of the last chapter and wish to remind the reader: It is people who make systems work—not technology, not management structures, and certainly not money, although each of these factors influences how, when, and if people will work at all.

NOTES

1. Wang Laboratories, Inc., *On Human Factors* (Lowell, Mass., 1983), p. 5.
2. Stephanie K. Walter, "How to Avoid the 'Coffee Clash,' " *Management Technology*, March 1984, p. 72.
3. See Resource Guide, Chapter 10.
4. William Atkinson, *Working at Home; Is It for You?* (Homewood, Ill.: Dow Jones-Irwin, 1985), p. 89.
5. Wang Laboratories, *On Human Factors*, pp. 8–9.
6. Richard Slatta, Ph.D., "The Problems and Challenges of the Computer-Commuter," *Link-Up*, June 1984, pp. 36–39.
7. William Atkinson, *The Psychology of Working at Home*, Telecommuting Technology Conference, 1985, Boulder, Colorado.
8. Evan Peele, "How to Make Telecommuting Work," *Personal Computing*, May 1982, pp. 38–40.
9. Atkinson, *Working at Home*, p. 111.
10. Ibid., p. 42.
11. Peelle, "How to Make," p. 39.
12. Wang Laboratories, *On Human Factors*, p. 8.
13. Atkinson, *The Psychology*.
14. Atkinson, *Working at Home*, p. 26.
15. Ivan Cutler, "Designing the Office around the Computer," *Personal Computing*, November 1984, p. 70.
16. Jeanne Woldenberg, "Telecommuting: No Workplace Like Home," *Words*, June–July 1984, pp. 24–27.

17. Verna Noel Jones, "Work Industry Sliding into Home Base," *Rocky Mountain News* (Denver), April 6, 1984, p. 51-W.

18. Peter K. Dallow, *Telecommuting in Fort Collins: A Case Study,* October 15, 1985.

19. Atkinson, *The Psychology.*

20. Patrick Honan, "Telecommuting: Will It Work for You?" *Computer Decisions,* June 15, 1984, pp. 88–92.

21. Arthur Miller, "Are You a Lousy Listener?" *Industry Week,* August 5, 1985, pp. 44–45.

22. Tom Peters and Nancy Austin, "A Passion for Excellence," *Fortune,* May 13, 1985, pp. 20–32.

23. Dr. Evan Peelle, "How to Make Telecommuting Work," *Personal Computing,* May 1982, p. 40.

24. Wang Laboratories, *On Human Factors,* p. 3.

25. Charles Steinfield, *The Nature of Electronic Mail Usage in Organizations: Purposes and Dimensions of Use* (abstract presented to the International Communication Association, San Francisco, California), May 1984.

26. Marilyn Webb, "Life in the Electronic Cottage," *Working Woman,* December 1983, pp. 106–109.

27. "Who Told You That?" (chart), *The Wall Street Journal,* May 23, 1985, p. 37.

28. Eric R. Chabrow, "Telecommuting: Managing the Remote Workplace," *InformationWEEK,* April 15, 1985, pp. 27–35.

29. Wang Laboratories, *On Human Factors,* p. 5.

30. Webb, "Life in the Electronic Cottage," pp. 106–109.

31. Mike Lewis, "If You Worked Here, You'd Be Home Now," *Nation's Business,* April 1984, pp. 50–52.

32. Peter F. Drucker, "A Prescription for Entrepreneurial Management," *Industry Week,* April 29, 1985, pp. 33–40.

33. Steinfield, *The Nature.*

34. Peelle, "How to Make," pp. 38–40.

35. Gerald M. Weinberg, "The Squabble over Sunlighting," *Management Technology,* March 1984, pp. 71–72.

36. Erik Eckholm, "Emotional Outbursts Punctuate Conversations by Computer," *New York Times,* October 2, 1984, p. 19.

37. "Smoking Policies" (graphic), *The Wall Street Journal,* September 3, 1985, p. 23.

38. Honan, "Telecommuting," pp. 88–92.

II

Telecommuting Settings, Technologies, Networks

This section of the book focuses on various:

1. Telecommuting settings.
2. Equipment used in those settings.
3. Networks that deliver information and communications to and from the telecommuter.
4. Future trends associated with telecommuting.

5

Telecommuting Settings

"HOME SMART HOME"

The telecommuter's immediate work environment is the workstation, its furnishings, and its surroudings. Together these elements and their ergonomic characteristics have considerable effect on the remote worker's performance. A recent National Bureau of Standards study cites the design of the workstation and its job features as being essential to VDT-based work. The study also points up the need to accommodate work areas to the physical differences and preferences of workers, and recommends the following:

- The individual should have some say in the design and selection of furnishings and equipment, as well as an opportunity to personalize the work space.
- Adjustable work surface height, keyboard level, document holder, and terminal screen.
- Chairs with high backrests and adjustable inclinations.
- Forearm/hand supports for keyboard work and movable keyboards.

Computer specialists from the Department of Computer Science, Concordia University, Montreal, Canada, are designing an "ideal" professional remote workstation for the home.[1] It is expected to provide a personal information system and a secure communications gateway for information, as well as to integrate all normal office functions for the telecommuter. To other office personnel who interact with the remote worker this facility will simply appear to be another office workstation.

All vital home management information will be readily accessible on the computer, and the device will be used to control appliances, energy, and resource consumption. In addition, the computer will enable the worker to access various information services such as consumer networks and banking.

Whether the employee works at company headquarters, in a work center, or in a home setting, office space design flexibility is especially important because manufacturers are currently producing equipment that reflects changing technology and upgrades the power and versatility of the worker. "Make it really flexible and easily changeable," advises a work space design specialist, because "some very responsible clients do change the configuration of workstations as many as 10 times per year if the work demands it".[2]

ELECTRICITY AND WIRING

Electricity, as it supports equipment in the automated office or home, is of special concern to the organization and telecommuter alike. Telecommuters should be cognizant of the issues and certain that their remote office has separate electrical circuits with sufficient capacity to support all computer-based equipment. See Table 5–1 for a comparison of the amount of power consumed by many household devices.

In the home setting, telecommuting equipment and power-intensive household devices such as refrigerators should be operated on different circuits. Otherwise, the remote worker may experience a "disk-head crash" or other computer failure. To safeguard the home computer sys-

TABLE 5-1   Consumption of Energy-Using Devices in the Home

Appliance/Device	Amount*	Cost†
Heating: Oil	750 to 1,500 gal.	$794 to 1,588
Gas	1,000 to 2,000 ccf	400 to 1,200
Electric	20,000 to 40,000 kwh	2,000 to 4,000
Air conditioner (window)	1 kwh/hr.	.10/hr.
Clothes dryer	960 kwh	96
Attic fan (summer months)	400 kwh	40
Window fan (summer months)	240 kwh	24
Refrigerator: (16 cf. frost free)	1,620 kwh	162
(8 cf. conventional)	360 kwh	36
Freezer: Chest type	1,100 kwh	110
Upright type	1,600 kwh	160
Lighting: Small house (5 rooms)	900 kwh	90
Medium house (7 rooms)	1,020 kwh	102
Large house (10 rooms)	1,200 kwh	120
Outdoor (5 hrs./evening, 100W)	180 kwh	18
Television: Black and white (180 hrs./mo.)	360 kwh	36
Color (180 hrs./mo.)	720 kwh	72
Washing machine	72 kwh	7
Water heater (electronic)‡	7,200 kwh	576

*Electrical use based on average homes, family of four, at $.10/kwh.
†Annual figures in dollars unless otherwise noted.
‡Water heater calculated at $.08/kwh.
SOURCE: *Communications Technology*, August 1985, p. 73.

tem, the user should install an inexpensive electrical spike filter. For those homes with many thousands of dollars invested in equipment, more sophisticated power protection is recommended. An uninterruptible power source (UPS), usually consisting of a bank of batteries that acts as a power source during power failure, can also protect the telecommuter's work.

Wiring systems must fulfill the following four primary criteria:

- Sufficient cable capacity to feed electronic and communications systems as well as other home devices.
- Reasonable cost for moving and changing power outlets should different equipment or location be desired.
- Economy and performance during its useful life cycle.
- A pleasing appearance, i.e., the uncluttered work environment, avoiding the wire mess look.

Because most telecommuting devices require wiring, people involved in remote work programs need to understand basic wiring options. Typical wiring systems include:

- Ceiling systems.
- Floor systems.
- Underfloor ducting.
- Floor drilling or "poke-through."
- Flat wire.

Table 5–2 compares building wiring options. It will be increasingly necessary to consider this issue in planning homes that are intended for telecommuters.

Static electricity is another issue that is as critical in the telecommuter's home as it is in the office. Static electricity occurs in two ways, through:

- Induction charging—Electrical fields radiate from the surface of materials such as styrofoam and cloth. In fact, even a shirt sleeve can generate a charge that is strong enough to erase computer chips or memory.
- Contact charging—In contact charging, two materials touch each other and then separate, causing one material to strip electrons from the other. This sometimes happens when one walks across a carpet and receives a shock on touching the doorknob. The danger is that while voltages are often in the 10,000 to 15,000 range before people notice an unpleasant or painful sensation that will alert them, it only takes 3,000 volts to disturb computer circuits.

Common sources of static electricity are:

- Waxed, painted, or varnished surfaces.
- Vinyl tile flooring.

TABLE 5-2 Trade-Offs among Alternative Wiring Systems

System	Positive	Negative
Poke-through	Low first cost No raceway	Expensive to make changes Disruptive of activities Fire safety precaution
Electrical cellular deck	Relatively flexible Easy to make changes	Expensive Service fittings sometimes protrude; unsightly, tripping hazard
Underfloor duct	Accessible No fire safety problem	Expensive Difficulty planning for needed flexibility
Raised access floor	Integration of electrical communications, HVAC Flexible	Expensive to install When frequent changes made, disruptive, hazardous Many spaces need conventional floors—ramps, stairs
Ceiling-based systems	Easy to integrate with furniture systems Flexible	High first cost Careful planning needed—intrusive
Flat cable	Short installation time Structural integrity of floor system kept Changes can be made readily Can be used under carpets	Cable must be shielded from punctures Careful installation needed to avoid unsightly cable ridges

SOURCE: Cross Information Company.

- Polystyrene (Styrofoam) materials.
- Finished wood or plastic-covered desks and chairs.
- Electrostatic copiers.
- Insufficient humidification.

To avoid static electricity problems, one must take precautions to ground electronic devices and the main computer system, avoid synthetic carpeting, and use antistatic mats under equipment. Installing ionizers (negative ion generators) is a relatively new method used to reduce static. Most offices that contain electronic devices are depleted of negative ions. Not only does equipment tend to become positively charged, but air conditioning ducts are positively charged, and neutralized negative ions occur naturally in the air.

Dust in the air, cigarette smoke, and various other pollutants attract static electricity as well as reduce the life of hardware, damage removable hard disks, and cause prematurely high error rates in all magnetic storage media. The sensitivity of central processing equipment means that it is essential to use an air filter or cleaner where such machinery is used extensively.

AIR SYSTEMS

When telecommuting takes place in an intelligent home, a variety of cooling requirements are critical. Although most equipment tolerates a wider range of thermal conditions than people do, concentrations of computing and other devices in an area generally produce a considerable amount of heat which can easily rise above suitable levels for the equipment. This may produce data errors or total equipment failure. In fact, some studies show that a computer terminal alone can generate 150 percent more heat than a human. Cooling is therefore important so that equipment failure is avoided.

It should be noted that a worker's performance is more closely related to the environmental temperature than is generally recognized and can be related to mood disturbance and aggression as well as to contrast sensitivity.[3] For most people, comfortable working temperatures are said to range between 68°F and 75°F, and acceptable humidity to range from 30 percent to 60 percent. Heating and air conditioning systems, and humidifying and dehumidifying systems can help maintain these comfort levels.[4]

When the telecommuter or organization purchases air conditioning equipment, workload is an important factor to consider. The time of day, day of the week, and season that the telecommuter works all contribute to the heat that air conditioning equipment must control. Depending on a worker's total air conditioning and associated electrical requirements, many power companies will offer assistance in planning and designing this aspect of the home office.

Because air conditioning systems, whether window or centralized, consume a substantial amount of electricity and are an additional cost, they must be included, along with ongoing utility costs, in assessing the "business case" for telecommuting.

Humidity and Pollution Control

Many devices used in the telecommuting setting are humidity sensitive and therefore require humidity control for their proper operation and for minimizing the detrimental effects of static electricity. If relative humidity is too high, moisture absorption and dimensional changes occur in paper that can cause paper handling devices to jam. Static results when humidity is too low.

Relative humidity and barometric pressure also have diverse effects on people, depressing some individuals and stimulating others. Certain people are also deeply affected by cigarette smoke levels and other atmospheric pollutants. Using a heat exchanger (indoor/outdoor) and an ionizer, ventilating the work area, and placing houseplants around the room will help overcome these problems.

In regard to indoor pollutants, James E. Woods, Ph.D., senior staff scientist for Honeywell's Corporate Physical Sciences Center, reported that people spend about 95 percent of their time indoors where pollution levels may be 10 times greater than those outside.[5] Indoor pollution caused by tobacco smoke, cooking, kerosene heaters, and cleaning solvents can create positive ions and decrease negative air ions.[6]

Cigarette smoke and ventilation are prime problems for both home and office employees. A sampling of Honeywell office workers who were polled concerning these problems complained that such pollutants caused them to:[7]

- Feel sleepy—56%.
- Have nasal congestion—45%.
- Experience eye irritation—41%.
- Have breathing difficulties—40%.
- Develop headaches—39%.

While Honeywell workers varied in their responses to questions regarding air quality, women (overall) believed air quality to be less satisfactory, possibly because they were more likely to work in fully enclosed offices and have less mobility than the men.

There are currently no mandatory standards set by the Environmental Protection Agency regarding concentrations of oxygen, carbon dioxide, gases, vapors, and radon. Therefore much more information must be gathered and studied regarding the adverse effects of pollutants and the threshold levels at which they affect health, as well as their cumulative effects after low-level, long-term exposure.

When noise pollution troubles the telecommuter, the situation can generally be masked by using white noise generators, fans, and background music. In a resonant room, rugging, thickly covered furniture, and woven hangings usually mute hollow sounds.

LIGHTING

Proper lighting is an important productivity issue to the telecommuter who spends many hours in front of a terminal. Screen readability is generally improved by designing the office to provide indirect lighting and by using individually controlled desk or workstation lighting. Control allows the user to modify lighting according to the time of day, mood, work activity, or other needs.

It also permits the user to increase lighting intensity for reading, writing, or working on detailed documents such as blueprints or circuit boards. Benefits are increased personal comfort and productivity as well as reduced potential for physical problems.

The source of adjustable light should always be outside, to the side or behind the user's line of vision, rather than in front and above it. In this way, light will not shine into the individual's eyes or create "veiled reflections," where bright light sources bounce off working surfaces and diminish one's sense of contrast. A light dimmer is considered helpful in controlling this effect. The computer should be placed at right angles to the light source so that the screen will not reflect it.

Lighting installation in the home as well as the office environment should support:

- The visual needs of those performing VDT and/or paper-based tasks.
- Controls that allow the individual to manage lighting for different needs.
- Easy maintenance.
- Avoidance of extreme brightness, contrasts, and glare.
- Integration of daylight into the overall lighting arrangement.
- A sense of personal office intimacy and delineation of both personal and common activity spaces.
- The architectural design of the home, e.g., its connection with the out-of-doors.
- Schemes that permit a reduction in energy consumption and heat load.

Systems that allow uplighting, task-ambient systems, and ceiling schemes with low strength light and reflectors designed to maximize diffusion of light should be considered by designers.

Professional-looking quarters generally give attention not only to lighting but also to color. Reds and yellows can be used to stimulate sedate people; greens and blues to relax more active individuals.[8]

Ergonomics, the study of the human-machine interface, may eventually become a legislated issue in America as it is in many European countries where there are laws regarding quality-of-life issues in the office. Some American cities and states are currently following suit by establishing no-smoking laws for public and private work places. California, among other states, is considering legislation pertaining to work-setting issues such as furniture construction, lighting, heating, cooling, static, and airborne pollutants.

WORKSTATIONS AND FURNITURE

Ergonomics is critical to the design of the effective office workstation, whether the site is company headquarters or the work-at-home environment. In an effort to produce ergonomic workstation standards, Public Works Canada Architectural and Building Sciences Department has developed a FUNctional and DIagnostic work enclosure called a FUNDI.

The purpose of developing the FUNDI is to determine the environmental and ergonomic requirements of traditional and new electronic offices as well as to learn how individuals cope with environmental stresses in the office setting.[9]

The FUNDI is a mobile, self-contained unit that rests on castors. It is flexible in that it allows users to adjust the internal environmental systems such as temperature, lighting, noise, ventilation, and air purification according to their preference. This monitoring system is combined with a functional work environment that provides the user with a range of choice that falls within current Canadian government standards for space layout and low-energy consumption. The unit also integrates data on tasks, the occupants' environment, and information on building facilities.

The Canadian goverment's interest in the automated office environment grew out of concern that industrial and office environments might degenerate as relatively new, computerized equipment, recently developed materials, and environmental systems came into use. Together, these elements are suspected of having an as yet unknown effect on human physiology and of creating stress that might lead workers to believe that their employers do not support their efforts to be productive.

The characteristics of a FUNDI unit include:

- Unit flexibility. The FUNDI comprises two hinged sections on castors. When fitted securely together, a small female can easily move the two sections under "worst case" conditions.
- Communications connections. Through an accordianlike structure, power, telephone, and data connections are maintained.
- Furniture flexibility. Fully adjustable VDT workstation (including adjustable footrest), work surface height, angle, and supplementary area (all of which can also be folded away) are available.
- Storage shelf, drawer, and fold-out table.
- Two skylights (one opens for air circulation) and a window to provide light.
- Ceiling of the shorter section raises.
- Infrared wall sensors detect a user leaving the unit and turn off the environmental system. When the user returns, the system reinstates itself to the last setting.
- Other environmental controls include: radiant heat panels, a fan, task and indirect background lighting. These are regulated by adjustable rheostats.
- Negative ion generator.
- Enhancements can be made by adding color, texture, and add-on components.

In the telecommuter's home setting, workstation flexibility might translate into something akin to the Murphy bed concept where the bed,

The FUNDI—a mobile, self-contained electronic office unit.

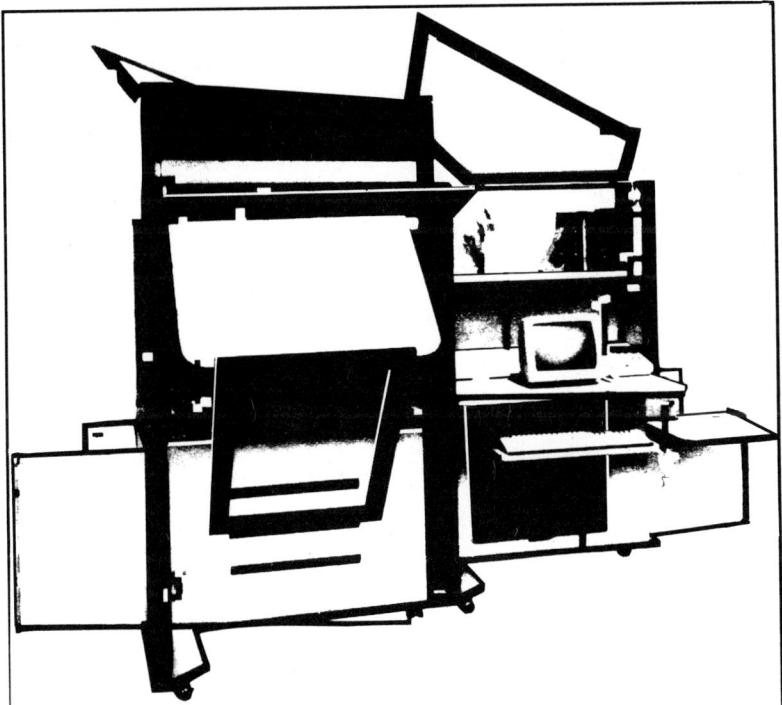

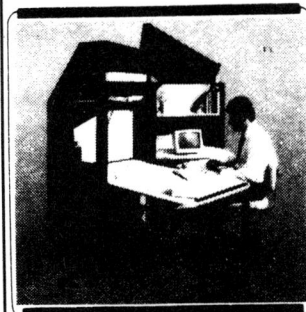

Courtesy Public Works Canada.

hinged to the floor, can be raised and folded upward into the wall. In some home environments, the workstation may need to be closed up in order to protect the contents from damage, children at play, or theft. Moreover, closing the work area at night might help the telecommuter realize that the business day is over and it is time to relax.

Workstation portability is of growing importance to companies, both in allowing equipment to be moved from one remote-work setting to another, and for reconfiguring the office setting. Movable walls, castors, and wire management systems are a particular advantage in changing the office design. Workstation wire management systems can organize wires according to their applications, for safety reasons, and aesthetics.

Furniture Design

The two key ingredients in furniture design today are:

1. Flexibility—the ability to adapt to changing needs.
2. Adjustability—the ability to tune the overall office environment into each worker's special needs.

See Figure 5–1.

New furniture designs help users cope with the problems of wire management and lighting. For example, there are modules designed as part of the furniture system (in box and circular form) where wire can be stored, and lighting systems that are part of the furniture design provide task and/or ambient lighting.

Furniture manufacturing is currently highly competitive, with manufacturers working to meet the users' changing requirements. To meet this challenge, furniture makers are producing:

- Desks and chairs that adjust to the different heights and sizes of users.
- "High-touch" furniture that comes in an endless selection of fabrics, shapes, colors, and textures.
- Furniture that is part of a system, so that only one or two pieces become obsolete as the user's needs change.
- Equipment and furniture that is designed and tested to guarantee an increase in user comfort.
- Furnishings with castors, or otherwise portable, for today's ever-changing office environment.

In the office and telecommuting environment, chairs should be supportive, adjustable, movable, and have padded arms. A number of innovative chair designs have come onto the market. Among them are the Norwegian "Balans" chairs. Their designs place the body at unconventional sitting angles that allow sitters to effortlessly hold their backs

FIGURE 5-1 The Electronic Communications Environment

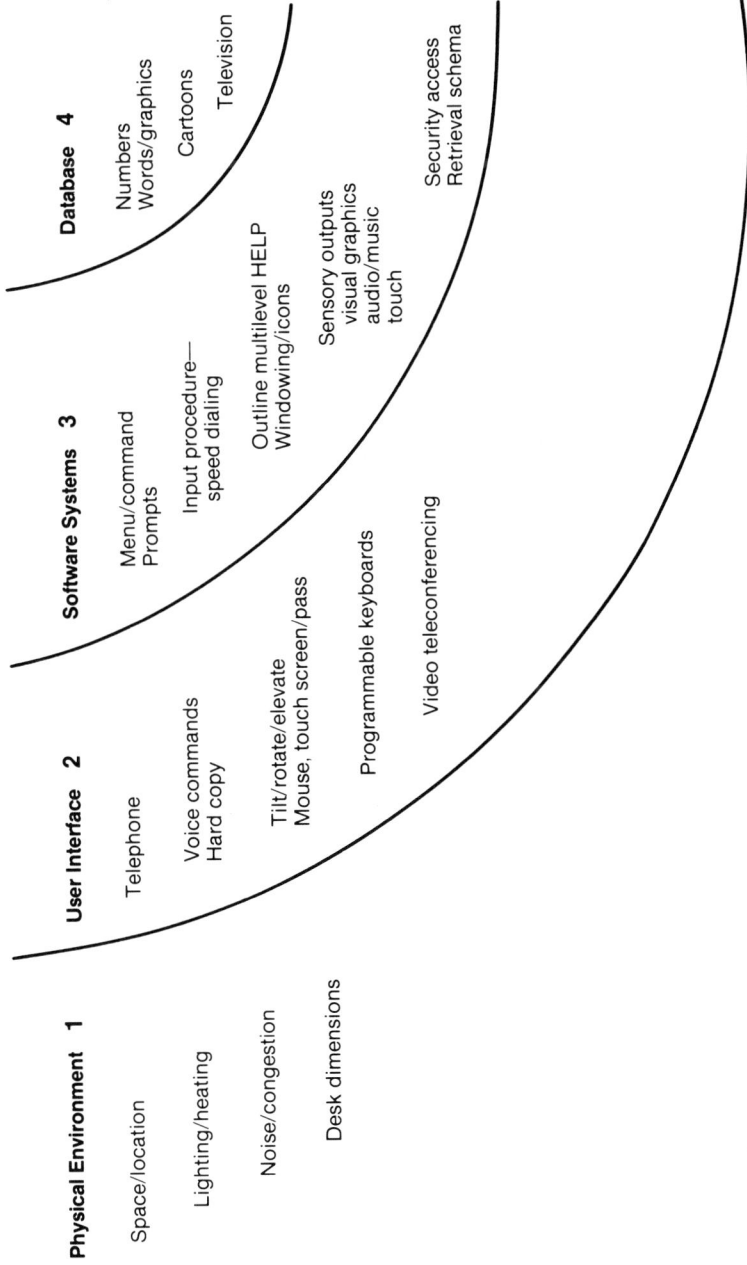

Physical Environment 1

Space/location

Lighting/heating

Noise/congestion

Desk dimensions

User Interface 2

Telephone

Voice commands
Hard copy

Tilt/rotate/elevate
Mouse, touch screen/pass

Programmable keyboards

Video teleconferencing

Software Systems 3

Menu/command
Prompts

Input procedure—
speed dialing

Outline multilevel HELP
Windowing/icons

Sensory outputs
visual graphics
audio/music
touch

Database 4

Numbers
Words/graphics

Cartoons

Television

Security access
Retrieval schema

SOURCE: Cross Information Company.

straight while remaining relaxed. The furniture on which the computer and keyboard rest should also have a matte finish so that light does not reflect off of it and into the worker's eyes.

The open office continues to be a very popular design concept, providing panel-hung walls, work surfaces, and shelves supported by dividers that reach above-normal standing height. Freestanding, modular equipment is appropriate for this type of arrangement as well as settings that have relocatable floor-to-ceiling walls.

Whereas furniture, lighting, communications distribution, and acoustics used to be separate areas of interest, industrial designers are now combining them. And furniture manufacturers are looking beyond the construction of traditional desks and storage to designing pieces that cope with the proliferation of computers.

"Igloos," an enclosure concept from Environetics International, Inc., are structural forms that can be set up within a large office area to provide the occupants with privacy, environmental, and some spacial control.[10] Such arrangements may be increasingly useful as future work areas become more self-sufficient and information is accessed by terminal. Igloo features include:

- Heating and cooling via an overhead duct.
- Cluster workstation arrangement set around a conference table.
- Full-featured workstations with printers, nine-section display screen, optical scanner, and equipment adjustability.
- Collapsible hood of woven glass or nylon embedded with lead. The hood adds density and reduces sound.
- Hood that has four accordian-type doors.
- Working and conference facilities for small meetings.

While the home may become a primary place for doing remote work, many employees will remain in the office. The emerging intelligent building creates better working conditions for office workers and, at the same time, provides better communications links with people working in the home and at other telecommuter locations.

THE "INTELLIGENT" OFFICE BUILDING SETTING

The intelligent or smart building is a key innovation that may have the single greatest impact on the office environment and telecommuting. Intelligent buildings are structures which generally include teleconferencing, telecommunications facilities, local area networks, energy management, and new information technologies. See Figure 5–2. In an increasing number of cases, these buildings are designed by computer-aided-design (CAD) systems that allow architects as well as tenants and occupants to have more say in the design of their working environment. (See the Resource Guide at the end of the book for current books regarding this topic.)

FIGURE 5-2 PBX/BAS Integration

PBX/BAS INTEGRATION

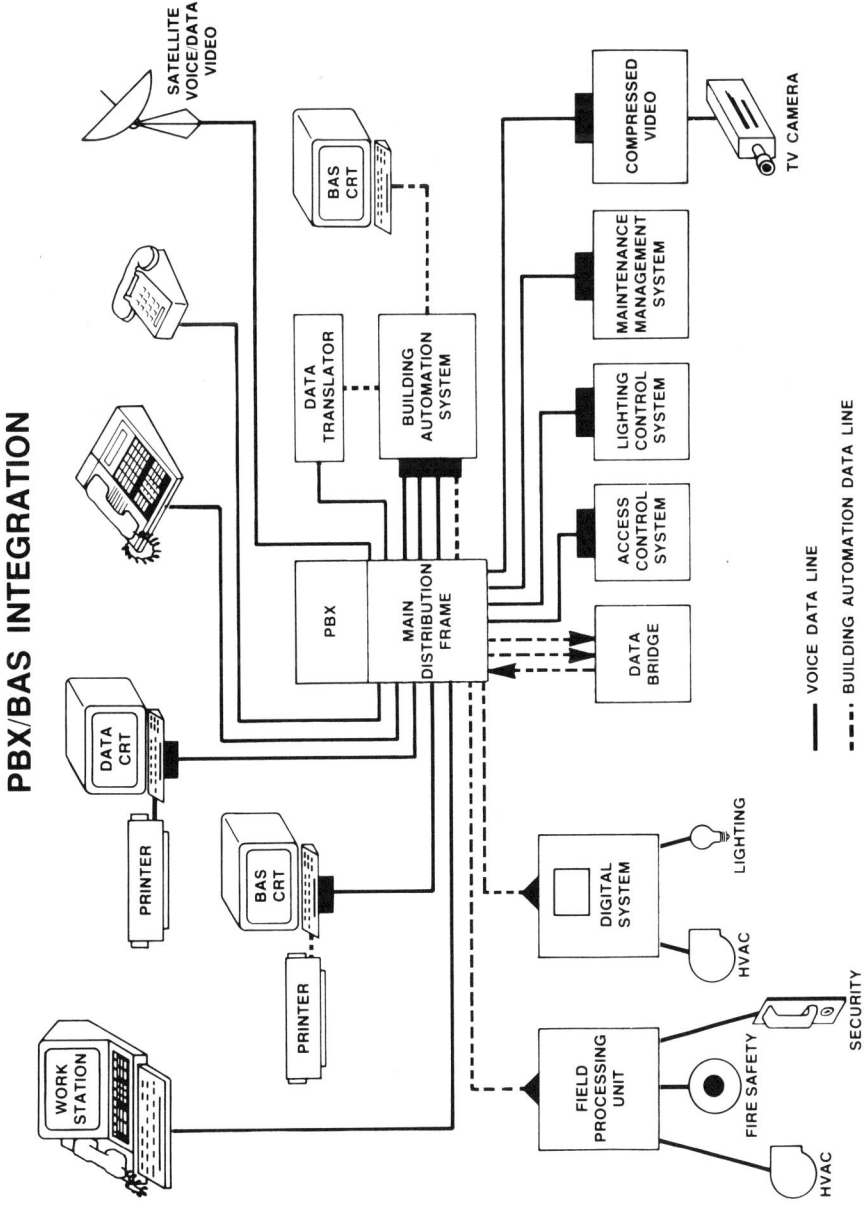

VOICE DATA LINE

BUILDING AUTOMATION DATA LINE

Courtesy: Johnson Controls.

The intelligent building is also referred to as telecommunications enhanced real estate, a multitenant telecommunications structure, and shared or enhanced tenant services. Much of the technology offered by intelligent structures, particularly in the area of networking and communications, supports the concept of the "transparent" office where workers perform tasks from any location and without regard to building walls.[11] The emergence of such structures is expected to heavily influence building developments in the 80s.

The intelligent building incorporates new office automation technology that allows employees to work more comfortably, communicate more effectively, and ultimately be more productive. However, companies that simply place "smart" office equipment within organizational walls can create barely controlled chaos—the state of affairs where equipment, desks, partitions, and people are being continually rearranged.

Because most automated companies use a complicated combination of terminals, telephones, and cables, more than one person and one department become involved in deciding what and who goes where. Consequently, organizations are coming to value building design features that simplify the automation of their offices. (Design changes can cover everything from good cable management to floor layout.)

Some of the many technological and social factors that affect intelligent building design include:

- Integration of leading-edge equipment. Most technology that will impact office design during the next 10 years already exists and is in use. But integration of such technology with existing office equipment and the rate at which the office worker will assimilate new systems into daily work life will be the major issues.
- The difficulty of assimilating information technology into office buildings and homes. The issue is whether investment in new technology truly helps workers improve job performance. If adding personal computers to an office configuration only aggravates a company problem, the office environment must be considered. Most office buildings are not designed to carry the environment load of new technology.
- Specifications for new intelligent buildings that are more stringent than is generally assumed. Data on building design that incorporates information technology is critical to planning. At present, office buildings are described in terms of their *construction* rather than in terms of the way they are *used*.

Future Building Issues

Management is expressing increasing concern that the office environment will be sufficiently suitable for both in-house and telecommuting employees. More specifically, companies are worrying that as they compete for personnel who are knowledgeable about new office technology, the environment will fail to attract and retain the best. As a result, some companies are exploring the ways in which telecommuters can be integrated into the workplace without requiring additional and often expensive office space.

Building owners are concerned that their need to reduce costs and more effectively manage building facilities (air, space, energy) will force the issue of incorporating new technology into building structures. It is in this area that new technology for telecommuters will provide greater interconnectivity between the serving office, the neighborhood work center, or the home.

In taking a long-range view of the office and telecommuting technology, the following "minitrends" should be considered: (1) home and neighborhood work centers are becoming economically viable work settings. (2) Companies that pay high rents in downtown business districts are considering how they can move certain less-essential office functions such as a data entry, telemarketing, and payroll functions to lower-rent outlying work centers.

In New York City, for example, some companies are moving data entry functions to New Jersey, Long Island, or overseas. Office furniture maker, Herman Miller, Inc., Zeeland, Michigan, is placing satellite offices that mirror the main office near residential neighborhoods.[12] These operations will need high-speed communications links to near or distant headquarters.

Technology is also changing the management structure of many companies. As a result some companies are increasingly delegating decision making to smaller and more autonomous business units, thereby centralizing some functions and decentralizing others. Certain businesses may find it desirable, for example, to centralize administrative services while decentralizing production. There is pressure on building designers to include options for more services that can benefit the telecommuter, including video teleconferencing, day care, health centers, jogging tracks, and even housing.

Information technology is affecting not only the size of office staffs and therefore office building requirements but also the ratio of skilled to unskilled employees. Consequently a different proportion of professional, managerial, and clerical personnel is evolving and influencing the numbers and types of telecommuters.

Greater accessibility to information via computer networking and better communications will change traditional, functional, and departmental boundaries. Whereas office workers needed to be together in large office buildings in the past, they can now be located thousands of miles apart and still feel close to one another.

Designers of telecommuter settings in the later 80s are merging utility and pleasure by combining the design elements of all places and times with current ergonomic principles. "The near-future daily environment," writer Charlie Haas claims, "may have walls whose backsides resemble the guts of fancy stereo tuners; miles of wires to service the profusion of audio (laser-played and graphic-equalized), video (cable-conveyed and VCR-immortalized), computers (no peripheral too peripheral to take its place on line), and remote controls that do everything, *including* windows."[13]

Currently, there are living quarters where a room's "soul," (a computer) is programmed to control the bed, turn on the coffee percolator, play wake-up music, raise window shades, regulate temperature, and match the lighting to the telecommuter's moods and activities. In one instance, the bathroom triples as a darkroom or bar.[14] With little or no renovation, bedroom alcoves, attic dormers, and large closets can provide working niches. And when the home traffic pattern permits, an infrequently used dining room can often double as a daytime working space.

As it is becoming possible to measure the benefits of implementing a telecommuter's home, one sees that work area design goes straight to the bottom line. This is also the reason that intelligent buildings are being electronically outfitted and the home and office settings will soon be part of one network.

Current trends indicate that work settings now include cars, trains, boats, and airplanes. And lap PCs permit people to work on a park bench or a golf course. Few technological limits to work settings exist, which means that people will try different possibilities to see how they work.

Conclusion

The remote-work setting, whether it is a work center, satellite office, boat, or home is opening new opportunities for companies and employees alike. For the company, telecommuting reduces the amount of office space, furniture, equipment, and technology required to support the worker. For the telecommuter, the remote setting may be more hospitable, convenient in time and place, and peaceful. This suggests that a balance will be required between people, machines, and furniture in a better work-setting relationship while companies pursue the often-elusive goal of increased productivity.

NOTES

1. P. Goyal and B. C. Desai, "A Personal and Remote-Work Station," *OAC 85 Conference Digest*, (Atlanta, Ga.: afips Press, 1985), pp. 25–31.

2. Charlie Haas, "Elements of Style," *Access* (*Newsweek* magazine), Fall 1984, pp. 76–81.

3. William Atkinson, *Working at Home; Is It for You?*, (Homewood, Ill.: Dow Jones-Irwin, 1985), p. 65.

4. Etienne Grandjean, *Ergonomics of the Home*, (London: Taylor & Frances, Ltd.; New York: John Wiley & Sons, Inc., 1973), p. 128.

5. "Again, Pollution Is a Major Concern; This Time It's the Indoor Variety," *Business Facilities*, July 1985, Critical Issues/Commentary column, p. 31.

6. William Atkinson, *The Psychology of Working at Home*, Telecommuting Technology Conference, 1985, Boulder, Colorado.

7. "Again, Pollution," *Business Facilities*.

8. Grandjean, "Ergonomics," pp. 242–44.

9. "Background on the FUNDI," *Public Works Canada*, Fall 1984, pp. 1–2.
    "Functionally Diagnosing the Office," *Public Works Canada*, July 10, 1985, pp. 1–4.
    "Fundi Project Summary," *Public Works Canada*, Summer 1985, pp. 1–2.

10. *Telecommuting Review: The Gordon Report* (newsletter), vol. 2, no. 2, pp. 12–13.

11. Francis Duffy et al., *The Orbit Study*, DEGW, England, 1984.
    Intelligent Buidings and Information Systems Report Boulder, Colo.: Cross Information Company, 1985.

12. Eric R. Chabrow, "Telecommuting: Managing the Remote Workplace," *InformationWEEK*, April 15, 1985, p. 32.

13. Haas, "Elements of Style."

14. Ibid.

6

Telecommuting Technologies

BACKGROUND

In any organization, goals of office efficiency and savings figure highly in choosing technologies. Companies that consider telecommuting must balance their potential costs as they apply to both "hard" dollar and "soft" dollar savings.

While people remain the most important factor in any telecommuting environment, the significant aspects of using telecommuting equipment include:

1. Easy installation and compatibility with the existing data processing equipment.
2. Choice of appropriate type of personal computers, from among the many available.
3. Security and archival issues, i.e., protecting both the users and the corporation from loss.
4. Communications technologies involved.
5. Networks to be used. They are the advanced "highways" where information travels. Current telecommunication systems are hardly more than "dirt roads" compared to the systems becoming available.

Telecommuting comprises a complex set of systems that requires understanding which tools to use and when. Furthermore, a company that employs telecommuters needs to plan for system expansion and increases in the number of people using the program as well as in the productivity of those who are already working remotely.

The integration of management and technology creates a balance between human resources and the computer "power tools" that assist people in accomplishing their tasks. The following equipment provides some of the advanced systems that are being given increasing communications capabilities. They include:

- Computers.
- Copiers/graphic scanners.
- Telephone systems.
- Networks.
- Smart desks and chairs.
- Electronic typewriters/word processors.
- Mail systems.
- Electronic file cabinets.
- Buildings/cars/homes.

The telecommuter's ability to reach out and communicate through these devices is one force behind his/her ability to work. The business community expects that, as international standards are applied in the computer and telecommunications industries, these systems will communicate via the same language.

The ability to communicate from many sites has spurred users on to using their equipment in increasingly more mobile work conditions. Equipment mobility, in turn, is affecting the physical needs and design of the office environment—possibly more dramatically than it affects the telecommuter. When office designers can truly address the increasing speed and efficiency of office work, as well as the psychological work environment, productivity may increase yet further.

For more than 300 years, offices have been created by interior designers and architects rather than by the people who understand office functions and procedures. This has resulted in work environments that are poorly suited to today's office tasks. It has also fueled general interest in telecommuting from other settings. In other words, technology now allows workers to relate to the office *electronically* rather than *physically*. As a result, people increasingly realize that technology can bring the job to them: They have the option of not moving where the job is.

It must be noted that technology often places pressure on managers and staffs, forcing them to:

- Improve productivity.
- Reduce "information float"—"time-to-decision cycles."
- Increase their effectiveness, efficiency, and spread.
- Decentralize business activities.
- Integrate information systems.

Each of these areas can be significantly improved through appropriately integrating and applying telecommuting programs. In addition, remote work can offer the "pressured" office worker a way to perform effectively at home instead of in a sometimes hostile setting. The technologies that may make the greatest impact will be those, like telecommuting, that *complement* rather than *replace* the office workplace.

TELECOMMUTING EQUIPMENT

Chapters 6 through 8 focus on the specific technologies available for telecommuting that are expected to support it. They are:

Chapter 6
- Telephone systems.
- Telephone answering services and machines.
- Personal computer networking.
    Modems.
    Local area networks.
- Telecopiers/facsimile equipment.

Chapter 7
- Electronic communications systems.
    Computer—electronic mail, bulletin boards, etc.
    Audio.
    Audio-graphic.
    Visual-graphic.
    Video.
    Slow scan.
    Radio.
    Voice mail.
    Cellular telephone.

Chapter 8
- Home automation systems.
- Robots.

TELEPHONE SYSTEMS

The telephone remains the most critical device for maintaining communication to and from the office and telecommuting environment. It allows people to stay in touch easily and conveniently. At the same time, one of the biggest problems about working in an office is telephone disruption. In order to allow quiet work time, some companies have instituted "no-calling" hours. These hours are usually from 8 A.M. to 9 A.M. The policy has limited value, however, because there is no way to stop people outside the company from calling in or employees from responding to corporate crises.

Telephones are an enigma for many people. They are so ubiquitous throughout the world that life would probably be impossible for most people without them. However, beyond using the telephone, few people have any idea of how they work, much less an understanding of the broader aspects of telecommunications technology.

Integrated Voice/Data Telephone

Courtesy Northern Telecom.

This section reviews telephone technology in terms that are useful to the telecommuter as well as to the office worker. This difference is important. If the office worker needs to arrange an audio conference with a telecommuter and other people located at distances, it is helpful to understand the PBX, CENTREX, or key system audio conferencing features which link them. In the future, telephone features will be sold according to their contribution to management productivity and effectiveness rather than on their apparent face value and amount of use, as they are now.

History

When the telephone was invented over 100 years ago, it was perceived as just another new technology in search of a customer. The mayor of New York was reported to have said that "every town will have at least one telephone." There are now over 180 million telephones in the United States and approximately 500 million throughout the world. In addition to the large external networks, business grew to need smaller telephone systems for internal communications. Today, business calls comprise about one third of internally generated calls. One third are local calls and the last one third are long distance connections.

TABLE 6-1 CENTREX versus PBX

When comparing CENTREX with the option of leasing or buying PBX equipment, today's business and residential user should realize that there is no one "right choice." There are only lower and higher risk versus economic and management choices. Among the factors to consider:

Issue	CENTREX	Key/PBX
Borrowing requirement*	No	Yes
Initial cost	Low	High
Ongoing cost	Uncertain	Rising
Maintenance:		
Response time	Immediate	2 to 24 hours
Cost	Included	$3 to $8 per set per month
Backup costs	Included	Optional
Space requirements	None	Expensive
Power requirements	None	Expensive
Multisite capability:		
Without additional equipment	Yes	No
Features:		
Currently	Few	Many
Future	More	Many more
Cost	High	High
Expandability	Easy	Difficult
Data transmission:		
Now	Low speed	High speed
Future	High speed	High speed
SMDR	Per station	Per system
Least-cost routing	No†	Yes
Office automation	No†	Yes
Moves and changes:		
Customer controlled	No†	Yes
Cost	Low	High

---

*Financial capability required to purchase.
†Presently restricted by regulatory authority.
SOURCE: Cross Information Company.

Key telephone systems (KTS) were small systems developed to provide for business calls. They initially allowed people to call internally via an intercom without tying up incoming and outgoing lines. They also allowed calls to be placed on hold and were used for "buzzing" people to alert them for calls.

In contrast, Private Branch Exchanges (PBX) were developed for offices with more than 50 telephones. The term *PBX* comes from the concept of a privately owned system located within a company (branch) premise switching system (exchange). Key telephone systems have evolved to the point where many experts consider them to be virtual PBX systems. This is because of the broad range of features and capabilities they have today and because many KTS have computers to process calls. Table 6–2 provides a feature comparison between key, PBX, and CENTREX telephone systems.

TABLE 6-2 Key Telephone, PBX, and CENTREX Feature
Comparisons

The CENTREX features were gathered from various Bell Operating
Company CENTREX offerings, although in some cases only a portion
of these features are actually offered. The chart is intended to
indicate CENTREX's feature-rich services, many of which are thought
to be found only in PBX systems.

	CENTREX	Key	PBX
Standard Features:			
Add-on conference	X	X	X
Call transfer—incoming	X	X	X
Consultation hold—incoming	X	X	X
Direct inward dialing	X	X	X
Direct outward dialing	X	X	X
Identified outward tolls	X	O	O
Station-to-station calling	X	X	X
Touch telephone dialing	X	X	X
Interception of bad calls	X	X	X
Consecutive station hunting	X	X	X
Automatic callback	X	X	X
Call forwarding	X	O	X
Call pickup	X	X	X
Additional call pickup group	X	—	X
Call waiting	X	X	X
Speed calling	X	X	X
Three-way calling	X	X	X
Off-premise locations	X	X	X
Call hold	X	X	X
Dialed conference	X	X	X
Distinctive ringing	X	X	X
Loudspeaker paging	X	X	X
Recorded telephone dictation	X	X	X
Reminder ring	X	X	X
Speed dialing	X	X	X
Alternate answering	X	X	X
Special functions:			
Moves and changes	X	X	X
Least-cost routing	—	X	X
Class of service restrictions	X	O	X
SMDR	—	X	X
Slim wire	X	X	X
Battery backup	X	O	O
Trunk queuing	—	X	X
Message waiting	—	F	X
Night service	—	X	X
Redundant CPU backup	X	O	O
Tenant service	—	O	O
Do not disturb	—	O	X

X = Yes, — = No, O = Optional, F = Future.
SOURCE: Cross Information Company.

CENTREX is a central office (CO) exchange switching system where the telephone company installs wires from the central office to each telephone. All CENTREX calls, even those within a single office, are connected via the CO. In many cases, the CO may be located miles away. In telecommuting applications, the telephones will not be in the office, but in homes or elsewhere. Citywide CENTREX can connect different COs and allow for simple dialing throughout the system.

A CENTREX system can give incoming callers the impression that people are at their desk when, in fact, they are at home. This type of arrangement has a number of advantages when telecommuters are dispersed throughout a city. In the future, CENTREX will provide additional important advantages geared for telecommuting.

There are certain factors that differentiate PBX and CENTREX switching technologies. Financing and service are the two most important. CENTREX users pay an installation fee and monthly charge per station. This is adjusted by the number of stations connected to the central office switch and the add-on features requested. These conditions permit smaller organizations, or organizations with limited funds, to avoid a heavy capital outlay for an on-premises switching center and for the associated office space.

Total system shutdowns are avoided by using CENTREX because each station is individually linked to the central office switch where a backup generator can provide service for the entire exchange when it is needed. The following are some advantages of using CENTREX:

- No capital expenditure.
- Easy expansion of the system to accommodate growth.
- Low installation costs.
- Space savings.
- No possibility of system failure.
- No charge for unwanted features.
- Technological and financial obsolescence are avoided.
- No need for uninterrupted power supply.

PBX and CENTREX systems were originally configured for the analog transmission found on traditional telephone lines, but today's trend is toward all digital traffic capability. New PBXs and COs offer digital transmission capability which permits very high-speed data transmission.

The Bell and other telephone operating companies have been making a concerted effort to replace or upgrade their central office switches to permit routine, high-speed data traffic. At the same time, fourth-generation PBXs (fully digital PBX systems) are expected to come onto the market during the next few years.

Advanced CENTREX

The evolving CENTREX picture is expected to offer so-called "Premium CENTREX," an enhanced version of today's CENTREX. There may, in fact, be many different versions of Premium CENTREX.

The following is a list of likely premium CENTREX features that will emerge:

- Enhanced station message detail recording (SMDR).
- Customer moves and changes.
- 9.6 to 56 kbps switched data transmission.
- Traffic management and reports.
- Electronic telephone instruments.

In addition, a number of new enhanced central office features will complement CENTREX. They include:

- Voice mail.
- Text/electronic mail.
- Energy management (home or office).
- Intelligent multitenant building management.
- Distributed CENTREX controllers.
- Alarms and telemetry—burglar.
- Data processing.
- Local area networks.
- Telephone answering services.

There are other possible CENTREX features as well. These include:

- Customer moves and changes.
- Least-cost routing.
- Local 800 intra-LATA service.
- Feature rearrangements.
- Message center activity.
- Customized station features.
- Interface to energy management and office automation.
- Security/traffic control systems.
- 9.6 to 56 kbps data/switched modem pooling.
- Citywide CENTREX (intra-LATA–FX–FCO).
- Multicity CENTREX and enhanced network management.
- Selective call forwarding—e.g., to the office.
- Distinctive ringing.
- Calling number display.
- Call rejection.
- Automated teller machine services.

Some features commonly found in all three systems are:

- Add-on conference, incoming and outgoing—Allows a telephone user to add another telephone station line into an existing conversation.
- Call transfer, incoming—Permits a station user to transfer any incoming call to another user within the station.
- Consultation hold, incoming—Allows the user to hold an incoming call through switchhook operation, and place a new call within the telephone for the purpose of consultation.
- Direct inward dialing—Incoming calls may be placed directly to the desired department or individual.
- Direct outward dialing—Allows station user to place calls without an attendant.
- Identified outward tolls—Identifies originating station for every outward long-distance call.
- Station-to-station calling or intercom—Allows any telephone station to call any other station by dialing four digits.
- Touch telephone dialing—Permits the use of touch-pad dialing.
- Interception of calls to unassigned number—Automatically intercepts calls placed to unassigned numbers within the telephone and directs the caller to a working number.
- Consecutive station hunting—Allows the incoming call to ring at another station when the dialed number is busy. This prevents lost calls.

Additional Telephone Features

- Automatic callback—A telephone user placing a call to a busy station is automatically called back when both stations become idle.
- Call forwarding—Automatically reroutes an incoming call to any desired telephone, within or outside of the telephone system.
- Call pickup—Allows any telephone user to answer a call ringing at another station.
- Additional call pickup group—A feature which allows for additional pickup groups. For example, the user can have some lines in one group, with the remaining lines assigned to another call pickup group.
- Call waiting—Alerts a station user on one call that another call is waiting. By using the switchhook, the user can hold the first call while answering the second. Using call waiting reduces the number of lines that may be needed.
- Speed calling/dialing—Enables user to place any call by dialing one or two digits. Up to either 6 or 30 numbers may be programmed in for speed or convenience dialing.

- Three-way calling—Allows the station user to establish a conference call with parties in or outside of the telephone system.
- Off-premise locations—Telephone stations can be installed at other business locations, allowing customers with multiple locations to have one-system convenience.
- Call hold—Allows the user to hold any call by depressing the switchhook and dialing a "hold code." The telephone set may then be used to place another call.
- Dialed conference—Enables the telephone station to establish a conference call involving up to six parties by dialing a conference calling access code.
- Distinctive ringing and call waiting tones—The station user determines the source of incoming calls by their different ringing or tone patterns.
- Loudspeaker paging—Station users can dial a code to access paging equipment.
- Recorded telephone dictation—Permits access to and control of customer-owned telephone dictating equipment.
- Reminder ring—Provides a half-second ring at stations utilizing call forwarding.
- Alternate answering—Automatically transfers outside callers getting a busy signal or no answer to another line for answering purposes.

In summary, there are numerous telephone features. Many of them will enhance the remote worker's ability to communicate effectively and efficiently with the office.

In the future, additional features to help remote workers are expected to include:

- Call rejection. If, for example, customers don't want to receive office calls while at home, they simply dial a two-digit code, along with the office number, and the system is programmed to reject those calls. The calls are then intercepted, where they are answered with a prerecorded message the customer has chosen.
- Customer-originated call tracing. If customers receive annoying or offensive calls, they simply hang up and dial a two-digit code. The originating telephone number of the last call is printed out at either the telecommunications company central office or the local police department. As with other features, customers are charged only for that time the service is used.

Most telephone companies plan to give customers more control over their telephones. New services will be conveniently provided in conjunction with a personal computer to control telecommunications services at one's home or office. They will permit video teleconferenc-

Touch-Screen Personal Computer

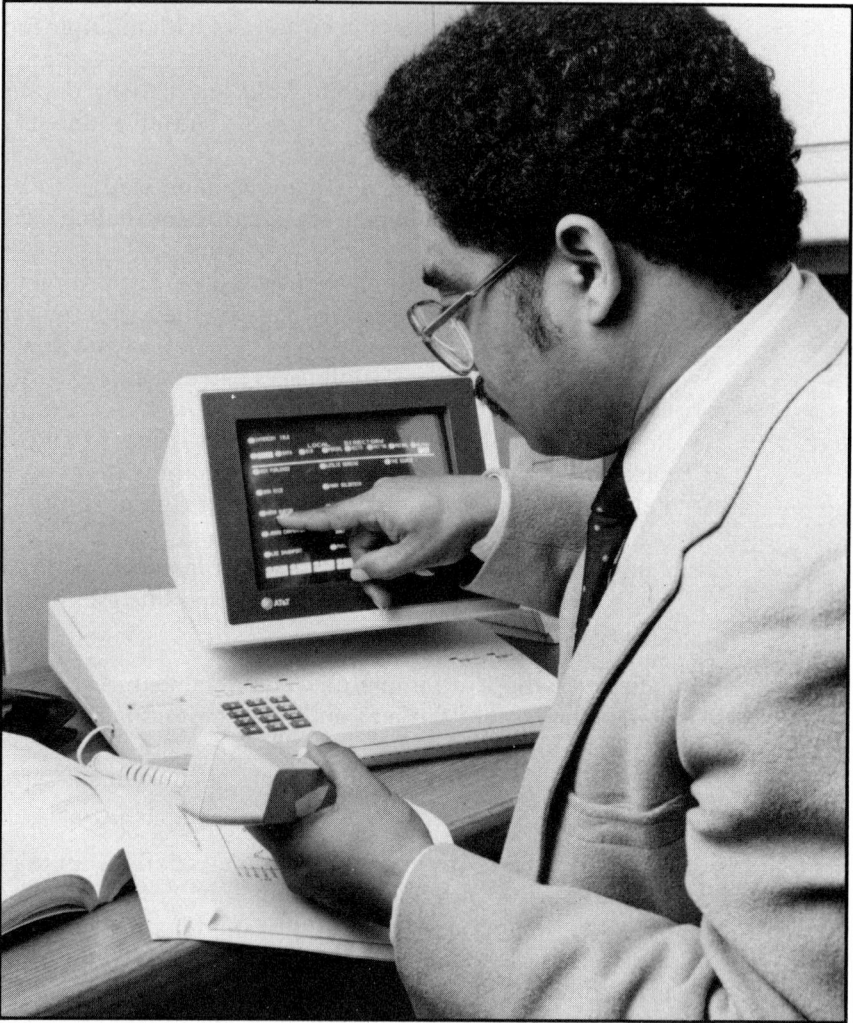

Photo courtesy AT&T.

ing, game playing, computer services, educational information, banking, and limitless information to reach the user cost effectively. In addition, telephones will be integrated with personal computers to facilitate many of these features.

Telephone Answering Services and Machines

The telephone answering machine is one of the most effective means for workers to control incoming telephone calls. Both telephone answering

TABLE 6-3 Comparison of Telephone Answering Service versus Answering Machine

Issue	Answering Service	Answering Machine
Initial cost	Low	Low
Ongoing cost	Increasing	None
Maintenance:	None	Some
Space requirements	None	Some
Power requirements	None	Some
Multisite capability	Yes	No
Features:		
Current	Few	Many
Future	More	Many more
Remote access	Yes	Optional
Expandability	Easy	Difficult
Calling records	Optional	None
Integration with PC	No	Yes
Multiperson voice mail	Yes	No

SOURCE: Cross Information Company.

machine and services have a number of advantages. Table 6–3 compares answering machines and services.

PERSONAL COMPUTER NETWORKING

The need to humanly gather, manipulate, organize, and distribute information has created the need for computer-communications networks. Thus, the development of communication via personal computer (using communications software) is considered one of the true driving forces behind telecommuting. By our estimates, more computing is currently taking place on PCs located on office desks than in vast computer centers. Moreover, we believe that in a short while more computing will take place in the home than in the office. For more information on this topic, readers are encouraged to read *Networking Personal Computers in Organizations* by James Weidlein and Thomas B. Cross (Dow Jones-Irwin).

The concept of a centralized mainframe computer as a communications control center for telecommuting remains and will probably continue to remain important for the foreseeable future. Consequently, we believe that microcomputer (PC) to mainframe connections will become a critical part of telecommuting network development. See Figure 6–1.

The basic networking elements of computer communications begin with the unintelligent, or "dumb," terminal (sending and receiving device) that is connected by cable to a modem (MOdulator-DEModulator). The modem is connected to a telephone line that runs to a controlling computer, as shown in Figure 6–2. This concept is generally oversimplified and can prove to be a complex process.

FIGURE 6-1

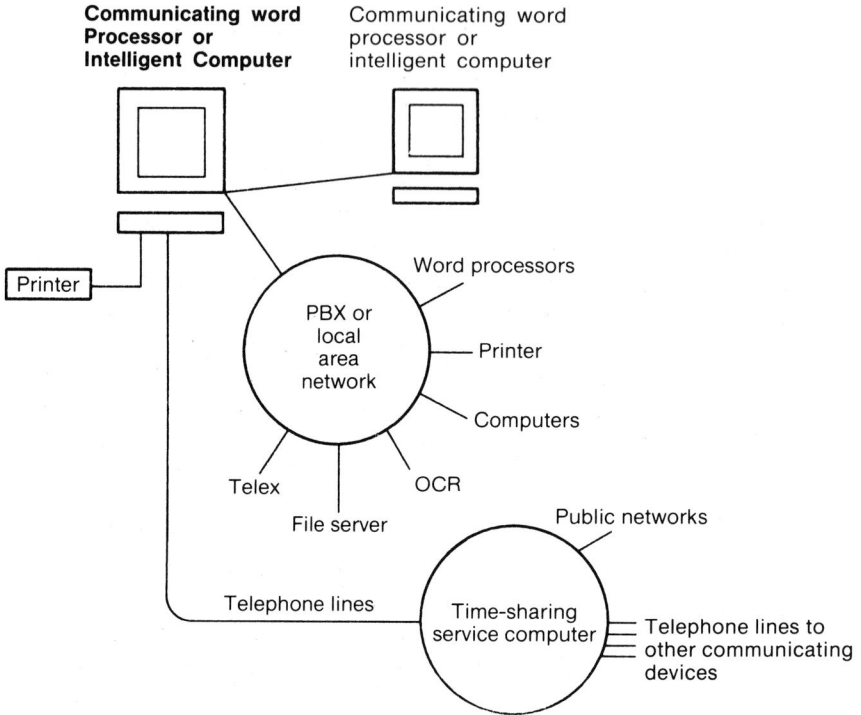

Communicating word Processor or Intelligent Computer

Communicating word processor or intelligent computer

Printer

Word processors

PBX or local area network

Printer

Computers

Telex

OCR

File server

Public networks

Telephone lines

Time-sharing service computer

Telephone lines to other communicating devices

Modems

One must *never* assume that computer communications will work. It is best to assume that a device will not work, until the manufacturer or vendor proves that it does (and may we add, with their money). A corollary to Murphy's Law might read: If there is any chance for a communications device to fail, it will. Overstated claims have resulted in a proliferation of lawsuits by users who did not get what was promised.

A typical problem is that not all modems communicate with the same language. Some letters or symbols are not understood by all models, and some modems have special characters which can disrupt software when PCs are used.

There are a number of common modem problems: (1) The terminal may be set at a different communication or bits per second speed than the modem. (2) The cable connecting the terminal to the modem may have a different pin wiring arrangement. (3) Or the modem may require a different plug than that provided by the telephone company or manu-

FIGURE 6-2 Connecting to a Computer Time-Sharing Service

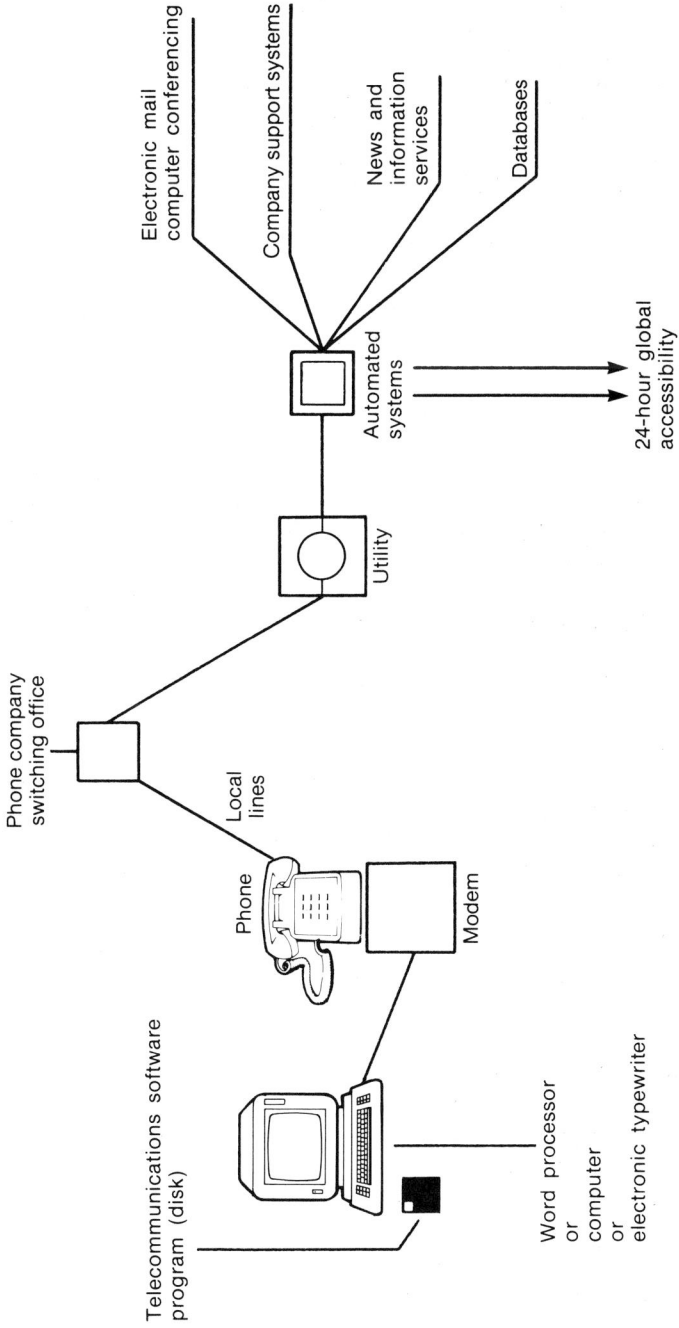

Telecommunications software
program (disk)

Word processor
or
computer
or
electronic typewriter

Phone

Modem

Local
lines

Phone company
switching office

Utility

Automated
systems

Electronic mail
computer conferencing

Company support systems

News and
information
services

Databases

24-hour global
accessibility

facturer. The person in charge of a telecommuting program must closely examine the equipment required for workable computer communications to ascertain if these are problems.

In considering modem and telephone connections between a telecommuter's terminal and an organization's mainframe computer, the speed at which information can travel between the two is frequently, and may continue to be, faster than that carried by common analog telephone lines today.

PCs can generally operate up to 19.2 kbps. However, most modems used for telephone-line communications today only carry 1,200 to 2,400 bps. That is but one tenth of the speed at which most PCs can communicate. It is important to recognize, however, that modem speeds over telephone lines have increased dramatically and will continue to do so in the future (to 56,000 bps and over 3 million bps). Speed limitations today are generally due to poor-quality telephone lines.

Speed is an important factor in telecommuting because it eases the individual's workload and gets information rapidly to and from the PC and mainframe computer. It should be noted, however, that the faster a modem transmits, the higher its price will be. Most corporate needs can be satisfied at 1,200 bps today, and 9,600 bps will suffice during the next few years.

Although a PC is connected to a modem, the devices will not necessarily interact. PCs and modems must have the appropriate communications software to do so and acknowledge this fact electronically. The basic function of communications software is to get information to and from another computer, whether that is a minicomputer, PC, mainframe, or other type.

Software can also control such internal PC communication as electronic mail, file maintenance, data integrity (making certain that data is accurately received), time programming (alarm clock functions for sending information when telephone rates are low), interfacing to spread sheets, and typesetting. Some software will also dial a telephone automatically and even store a variety of telephone numbers so the user doesn't have to remember them.

Once communications software is installed and a PC and modem recognize each other and their functions, electronic communications can be attempted. Unfortunately, this does not mean that the computer will talk to other PCs, or that devices will understand one another when they do connect.

The next step is to connect the computer to the modem and telephone line and make the computer telephone call. The telephone line has certain characteristics as well as modular interface jacks that can impact the communications being sent by the PC. In certain situations, noise will garble transmission, making it unintelligible. Other factors can also cause serious signal problems. Satellite telephone lines used by many of the long-distance common carriers have delays which can hamper the modem interaction.

A number of functions will have taken place when a PC has telephoned the distant computer. Most of them relate to acknowledging the existence of the sending terminal. For example, when a corporation's mainframe computer understands incoming signals, it will respond with appropriate messages, usually acknowledging terminal type (PC, dumb terminal, etc.), user name, and password.

This procedure has become difficult and tedious because most computer center managers have been cautioned against intruders (often called hackers) and demand increased security procedures. Some communications software packages can perform many of these connecting and password functions. After connections are made, telecommuting tasks can be sent and received. If problems ensue, the system will need to be tested, monitored, and adjusted.

Local Area Networks

Local area networks or LANs, sometimes called departmental PC networks, are a new technology that provides linkage between office devices and a gateway for incoming information. See Figure 6–3. Once communication passes through the LAN network gateway to the mainframe computer, the user can access files, run programs, send and receive mail, and enjoy other privileges granted by the system manager.

The concept of personal computer communications takes on different meanings when used with local area networks (LANs). For example, a LAN may interconnect office PCs and act as a gateway to the telephone network, other offices, a work center, or home setting. Telecommuters may call in to the LAN from their homes via PCs or dumb terminals as well. See Figure 6–4.

Once again, organizations should never assume communication devices will work. They should:

- Require vendors to prove their claims at the vendor's own risk and expense.
- Check vendor references.
- Get a second opinion from other users or a consultant.
- Develop a business case, if only a short one, to prove that there is a cost benefit.
- Try the system and test the software.
- Expand the system incrementally. Systems often do not work when they are enlarged.
- Read implementation suggestions in Chapter 4.

FIGURE 6-3 The LAN Ties the Office Together

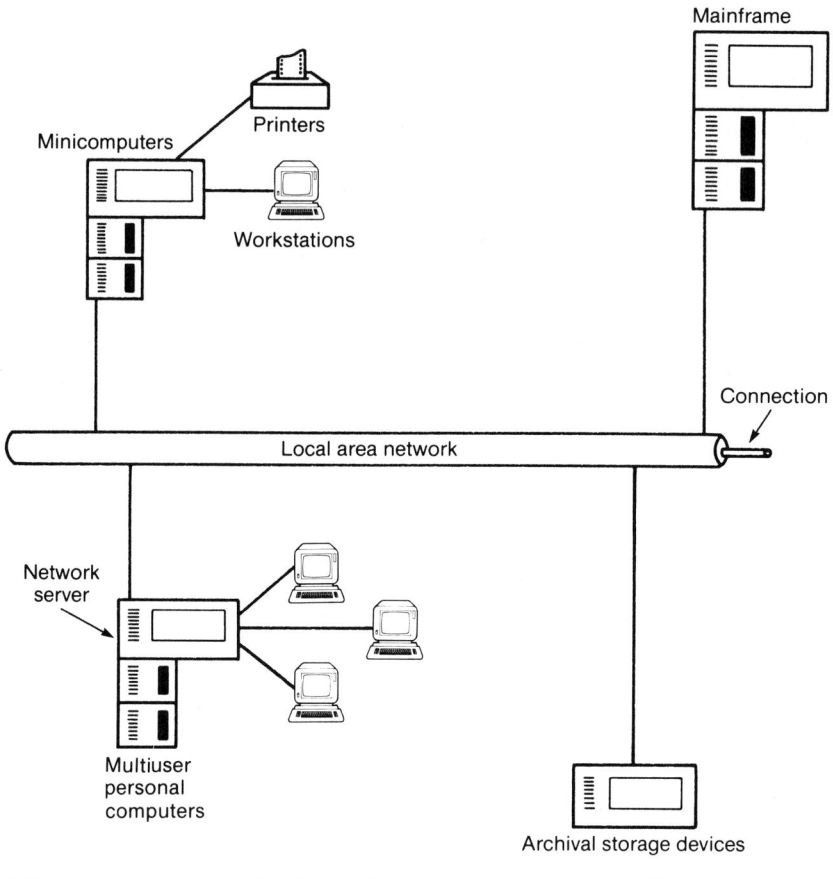

TELECOPIERS OR FACSIMILE SYSTEMS

Telecopiers or facsimile (FAX) provide an alternative to the postal system or carriers for the rapid transfer of hard-copy material at a reasonable cost. A telecopier can transmit the likeness of a document over telephone lines across the town or country in a matter of minutes. See Figure 6–5. The original is fed into the equipment, where it is encoded for transmission, processed, and returned. The operation is similar to photocopying. Telecopy, or facsimile, is an ideal method for transmitting numerical or financial data, drawings, and other material.

Telecopiers have evolved from machines that took six minutes to send one murky page, to equipment which transmits a letter-size page in 15 seconds or less. This speed reduces the cost of the associated tele-

FIGURE 6-4 LAN Typologies

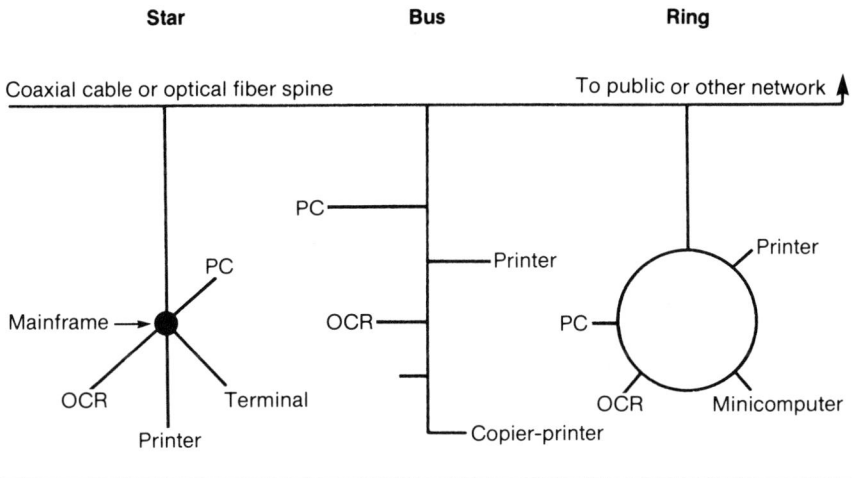

Star	Bus	Ring

Coaxial cable or optical fiber spine

To public or other network

Star:
- Mainframe
- PC
- PC
- OCR
- Terminal
- Printer

Bus:
- PC
- Printer
- OCR
- Copier-printer

Ring:
- Printer
- PC
- OCR
- Minicomputer

phone call. It also permits the user to transmit many-paged documents in a short period at a nominal cost.

Telecommuters can also use telecopying equipment within the organization to transmit documents among offices. This enables them to utilize the expertise of individuals at different sites to prepare a project or study without the delays associated with mailing documents between offices, or the expense of express delivery services.

Presently there are four major internationally approved facsimile system groupings. They transmit in different ways and at varying speeds:

Group 1   Four to six minutes per page to transmit.
Group 2   Three minutes per page to transmit.

FIGURE 6-5 Facsimile (FAX)

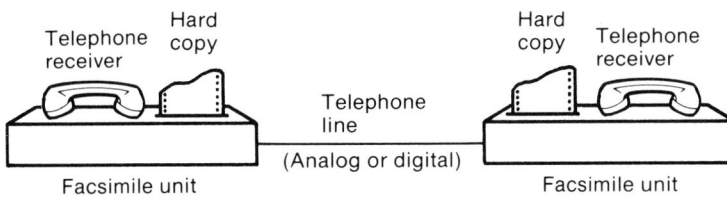

Telephone receiver · Hard copy · Telephone line (Analog or digital) · Hard copy · Telephone receiver

Facsimile unit   Facsimile unit

Group 3   Less than one minute per page—analog.
Group 4   Less than one minute per page—digital.

Developments in facsimile are moving at an ever-increasing rate. The driving factor is the need to reduce transmission costs paid to the long-distance communications carrier. Facsimile has important potential as a service offering for the telecommuting environment. In considering the purchase of facsimile equipment, an organization should evaluate the number of pages that can be transmitted per minute. The cost of operating a six-minute facsimile system as compared to a newer Group 3 or 4 subminute system may justify the purchase of the newer equipment.

Most major facsimile vendors have expanded their services to allow their equipment to interface with word processors or other computer systems. A growing number of express mail carriers also provide facsimile transmission as part of their overall package transport business. This interface capability facilitates the transmission of clients' lengthy documents and reduces equipment expenses for firms with light telecopy requirements.

The desired features of a telecopier are digital transmission, white line skipping, and automatic send and receive capabilities. Digital transmission and white line skipping contribute to rapid transmission speeds. Automatic send and receive capabilities allow even lengthy documents to be sent and received at unattended terminals.

The copier is another office machine that has recently been modified to handle communications. Intelligent copiers can now receive data on a computer via a direct-connection basis or by magnetic tape. Multiple copies can then be made automatically, thereby saving operating and printing time. By connecting a personal computer and modem to them, these intelligent copiers can communicate unattended, sending mail, graphics, and entire documents whenever necessary, including night hours when communications rates are lowest. Optical character readers also allow information to be gathered. See Figure 6–6. In the future,

FIGURE 6-6   Optical Character Recognition System (OCR)

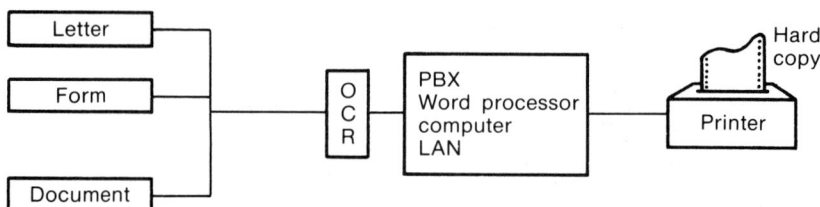

these devices will be able to transmit material to other kinds of equipment, combining copier and telecopier functions. "White" marker boards provide yet another means by which to transmit hand-drawn images from one location to another.

Conclusion

Telephones are the basic tools that allow the telecommuter to keep in touch with and/or otherwise be close to the office and co-workers. Besides telephones, there are numerous emerging technologies which allow the telecommuter greater flexibility in working remotely. These are discussed in the next chapter.

7

Electronic Communications Systems

BACKGROUND

Until the early 1980s, management worldwide spent hundreds of millions of dollars to improve factory productivity while virtually ignoring the office environment. As a result of this oversight, the cost of running offices has steadily consumed a larger part of companies' operating budgets. Three factors have made the emphasis on improving the office environment important:

1. Skyrocketing cost of air travel, hotel and motel accommodations, car rentals, and management salaries.
2. Increasing importance of reducing information float and decision-making time.
3. Practical new technology that enhances communication through teleconferencing.

In an effort to reduce rising office costs, organizations are attempting to streamline their communications systems, the most costly part of operating an office, and assert better control over the internal flow of information. A number of electronic communications systems support these efforts as well as provide an effective means for keeping telecommuters in touch with the office and other colleagues. They enable users to communicate from any distance, using text, graphics, audio (voice), video, or music.

As a result teleconferencing systems have become an increasingly important supplement to corporate communications systems. Most people audio teleconference everyday when they use the telephone. Organizations that bring together three or more telecommuters via teleconference can save precious management time and be far more productive.

IBM, J. C. Penney Company, Inc., and other organizations are beginning to use teleconferencing as a competitive tool that brings more people into contact, reduces decision-making time, and coordinates projects effectively.

Teleconferencing systems enable two or more people at various locations—including telecommuters—to communicate electronically. Depending on the type of system used, they can exchange information and examine drawings, plans, or sketches without interrupting their work schedules or paying for costly travel. Teleconference participants have also found that such communications systems can:

- Greatly reduce the need for travel.
- Break the habit of holding face-to-face meetings, thus accomplishing much more in less time.
- Allow people to deal with issues as they arise rather than postponing them until a scheduled meeting. This increases the speed of decision making and improves organizational efficiency.
- Shorten meetings and thereby increase efficiency.
- Provide a convenient, effective form of communications.
- Improve the quality, quantity, and speed of management feedback.
- Increase organizational loyalty and enhance job satisfaction by allowing many more people to participate in decision making.

People who stick with a teleconferencing system are known to be more committed and effective workers. Nonetheless, teleconferencing does not pretend to replace face-to-face meetings but rather to supplement them. It can, to a degree, provide the "closeness" of face-to-face meetings while allowing participants to locate where they choose and, in some cases, to join a conference when they want. Most participants agree that the conference manager is a prime factor in the success of any meeting, teleconferencing or face-to-face. See Figure 7–1.

Applied software systems for salespeople to use in working remotely have been calculated to increase the employee's selling portion of the day by as much as 50 percent.[1] These systems cover marketing, communications, and management areas, prospect and client tracking, training prospect analysis, managing sales force, customer letters, reports, sales management training, and organizing. Programs for the salesperson working remotely can also cover all office functions, including telephone messaging via electronic mail and computer teleconferencing.

For salespeople, lap-top units appear most effective in:

1. Allowing people to be in the office less time and on the road longer (communicating with the office via audio and computer teleconferencing). In this way, they do not need to return to the office early to finish their tasks.

FIGURE 7-1 Hierarchy of Meeting Formats

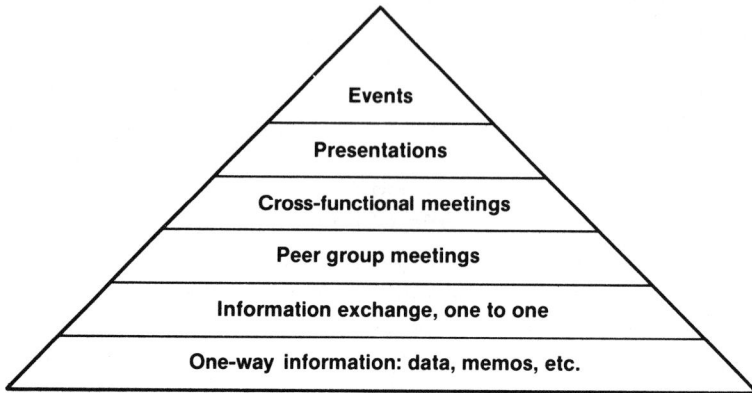

SOURCE: Courtesy Keiper Associates, Inc.

2. Enabling them to close deals on the spot instead of waiting for information, thus reducing cancellations by as much as 90 percent.
3. Improving the salesperson's image by having correct and extensive amounts of information available (accessing the corporate computer remotely).[2]

Comprehensive management studies show that executives and managers spend most of their time communicating and very little time doing desk work. By altering the means of communicating and substituting electronic for face-to-face meetings, organizations can save a considerable amount of money.

Recent studies indicate that less than half of all communications require face-to-face meetings. Topics discussed in meetings can often be handled in less costly ways. In fact, once most meetings and conferences are stripped of small tasks and formalities, their useful duration is usually less than one-half hour. Furthermore, one third of these meetings appear to be for the sole purpose of exchanging information rather than for decision making.

Teleconferencing is also a viable supplement for face-to-face meetings because it allows an easy exchange of information without the expensive, time-wasting formalities of traditional corporate meetings. However, telecommuting does generally require face-to-face meetings to reinforce and validate communications and prevent misunderstandings.

Corporate needs for teleconferencing are as varied in today's high-cost, information-based environment as the media and applications that

Problems Typically Encountered in Face to Face Meetings

Courtesy Cross Information Company.

have been developed to meet them. Typically, teleconferencing answers the need for:

- Frequent communication between remote sites—homes, offices, hotels.
- Communication between company departments—engineering to management to marketing.
- Business meetings in hard-to-reach locations.
- Avoiding high travel or telephone communications costs.

For more information on the subject, we suggest reading *Teleconferencing: Linking People Together Electronically,* by Kathleen Kelleher and Thomas B. Cross (Prentice-Hall).

The overall experience of people who have tied teleconferencing into a telecommuting environment has been encouraging. Participants agree that teleconferencing has a very positive impact on the work environment and on "management velocity," a term used to describe the speed and effectiveness with which management solves problems. Teleconferencing also provides the privacy and quiet that knowledge workers require in performing tasks.

Experience suggests that the home is a better working environment for activities that require concentration, while the office is better for communications, meetings, and social activities. Teleconferencing technologies allow telecommuters the best of both worlds—to work remotely, yet to be effective in the office.

ELECTRONIC COMMUNICATIONS SYSTEMS

The following section is an overview of the different forms of teleconferencing currently being used for telecommuting or remote-work applications. They are:

1. Computer (including electronic mail, bulletin boards, notepads, and conferencing features).
2. Audio.
3. Audio-graphic.
4. Visual-graphic.
5. Video.
6. Slow-scan.
7. Voice mail.
8. Radio.
9. Cellular telephone.

Communications via Computer

This is generally called computer conferencing, computer-aided communications (CAC), or computer-aided networking (CAN), and allows people in different locations to conduct an ongoing meeting using video display terminals (CRTs), personal computers, or other computer systems. An electronic message system records the telecommuter's communications, providing a verbatim log of the meeting. Each person may access, read, and respond to these communications, regardless of whether other participants are communicating simultaneously.

Computer teleconferencing provides an asynchronous (nonreal time or store-and-forward) method of participating that offers extraordinary flexibility. The technique has proven to be highly effective for managing ongoing project activities, and for communicating when participants travel frequently or are located in different time zones. Because telecommuters do not have to be in their homes or at their desks at a particular time to computer conference, time does not restrict their participation. In contrast, when face-to-face meetings are planned, travel and other meetings often cause scheduling delays and a reduced interest in them.

The following are some other key advantages of computer teleconferencing systems:

- It is generally the lowest cost of the teleconferencing technologies.
- It can use creative software systems to develop models for better decision making.
- It provides a long-term electronic filing system that enables the user to access large files which can be easily and rapidly retrieved.
- No special computer terminal training is required because on-line "help" and training is available.
- Communication is improved among managers. The network becomes a "place" in the thought processes of those who are connected via computer communications.
- Turn-around time is reduced on urgent decisions and actions. (In one test using computer conferencing, decision time was reduced from one to two weeks to one to two days, in many cases.)
- The number of interruptions from telephone calls is reduced.
- People are never late for a meeting. They use the system to reach places and people when it is convenient.
- The user is able to organize and reorganize messages and information for the most logical presentation. The discipline of putting thoughts into writing before communicating them improves the quality of communications.
- Tension among managers and telecommuters is decreased. Telecommuters are always connected to the office and never out of touch.
- The system produces copies.
- Techniques for solving problems are more effective because several people can review pertinent information.
- The system log can inform telecommuters on the status of their projects, from the beginning to the end. People can enter the process at any point and have full documentation to evaluate the process.
- Training of new staff members is accomplished at a lower cost per student. Current staff members do not lose time away from the job.
- Management/employee relations are improved for those affected by new company policy documents when information is presented at successive stages of development.

- Scheduling problems are eliminated because participants interact at their own pace and convenience. This allows the telecommuter to think through an issue before responding and provides a more fruitful, higher quality of deliberation.
- Some computer teleconference systems can operate in "real time," although they are rarely used this way except for management or emergency crises. This option influences other patterns of usage.
- Discussions can be held within larger meetings. Groups can convene electronically, then decide to meet or break up into committees of various sizes.
- Groups can consult easily and effectively for medical diagnoses. The doctor sends a description of the problem to other doctors who can then consider it and respond with their recommendations.

Far from discouraging personal participation, a computer teleconferencing system encourages users to write their ideas as they occur. This gives them the time for reflection that might not be available during face-to-face meetings. When computer teleconferencing is used to prepare for face-to-face meetings, background issues and trivia can be dealt with beforehand, so that people are ready to make decisions or move forward to more difficult questions.

A subtle yet key issue in most face-to-face or real-time meetings is that they move in a step-by-step, linear fashion taking up a great deal of time. Each speaker must wait until the previous one has finished before making a presentation. Thus, if there are 10 people and each one is allowed 5 minutes, the meeting must last a minimum of 50 minutes. And additional time is required for asking questions and responding. See Table 7–1 for advantages.

In a computer teleconference, each person presents material simultaneously and people respond when they choose, without interfering with others. This opens up a myriad of communications patterns unavailable by any other communications technology. Moreover, in crisis situations, when time is of the essence and problems are crucial, participants can interact, evaluate material, and make good decisions fast.

TABLE 7–1 Key Advantages of Computer Teleconferencing

- No time restrictions—"never late for a meeting."
- No geographical restrictions—"always there."
- Low cost.
- Self-documenting and filing systems—"electronic footprints."
- Self-pacing and training—"on-line help and training."
- Convenient participation—"on the road."
- No acting skills required.

FIGURE 7-2 Communicating during Computer Teleconferences

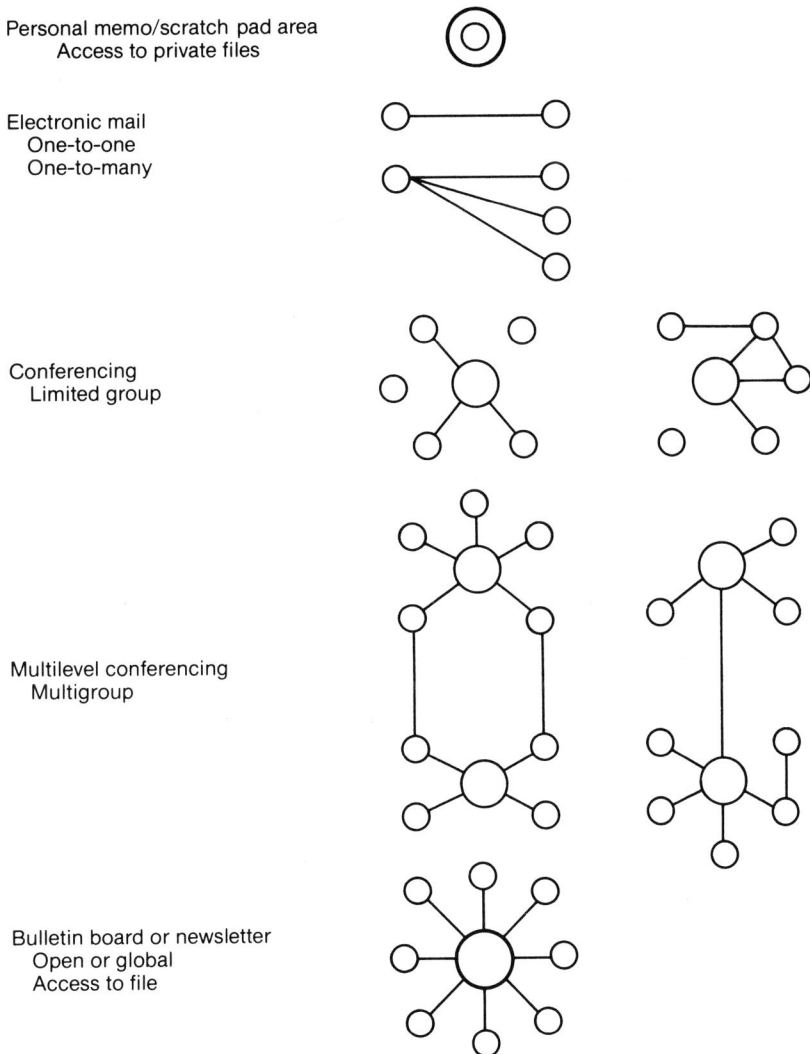

Personal memo/scratch pad area
    Access to private files

Electronic mail
    One-to-one
    One-to-many

Conferencing
    Limited group

Multilevel conferencing
    Multigroup

Bulletin board or newsletter
    Open or global
    Access to file

SOURCE: Courtesy Cross Information Company.

Options in communicating by computer. In most electronic communication systems, there are four basic levels of involvement, access, or use. See Figure 7–2. They are:

1. Bulletin board—open or global access.
2. Electronic mail—one-to-one communication.

3. Memo or notepad—private work area.
4. Conferencing—controlled or group access.

Bulletin board features. Bulletin board access is generally available to all users. While participation in conferences is limited to invitation or selection, the bulletin board is a useful place for announcements of a general or organizationwide nature. These might include job postings, club meetings, and management policies and procedures.

Business organizations have grown to realize that electronic bulletin board systems (BBSs) are not just telecommuter toys. They can be used in many ways and to considerable advantage in business by people at all organizational levels—within the company and outside of it—including telecommuters. See Figure 7–3.

In fact, the BBS can serve as a place to post such items as:

- Company policy notices.
- Work-related data.
- Organization newsletter.
- Personal messages.
- Technical documentation ready for review.
- Department inventory (number of products stored, shipped each month, on order, etc.).
- Telephone answering service (some systems accommodate lengthy correspondence).

Depending on the sophistication of the software that manages the system (electronic mail or computer teleconferencing), the BBS can provide a public message exchange, a private electronic mailbox and filing system for users, and a real-time communications system. Where firms have incompatible computers, employees can use BBSs provided by companies like the Reader's Digest Source or Compuserve's BBS. The BBS can replace the secretarial functions of taking dictation, copying, and sending messages by allowing the manager to compose, type, and send a message at one time, automatically and instantly.

Such systems have their limitations, however. For example, some limit messages to a maximum of 500 characters. "Junk mail," run-on letters, embarrassing language or "flaming," and filibustering can become problems. In a business situation, good BBS managers will review messages and information requests often and on a regular basis, monitoring for problems and helping employees improve their communications skills.[3]

Operating a BBS requires a personal computer, modem, telephone line, and the appropriate software. Among the popular software pack-

FIGURE 7-3   Teleconferencing Communications Options

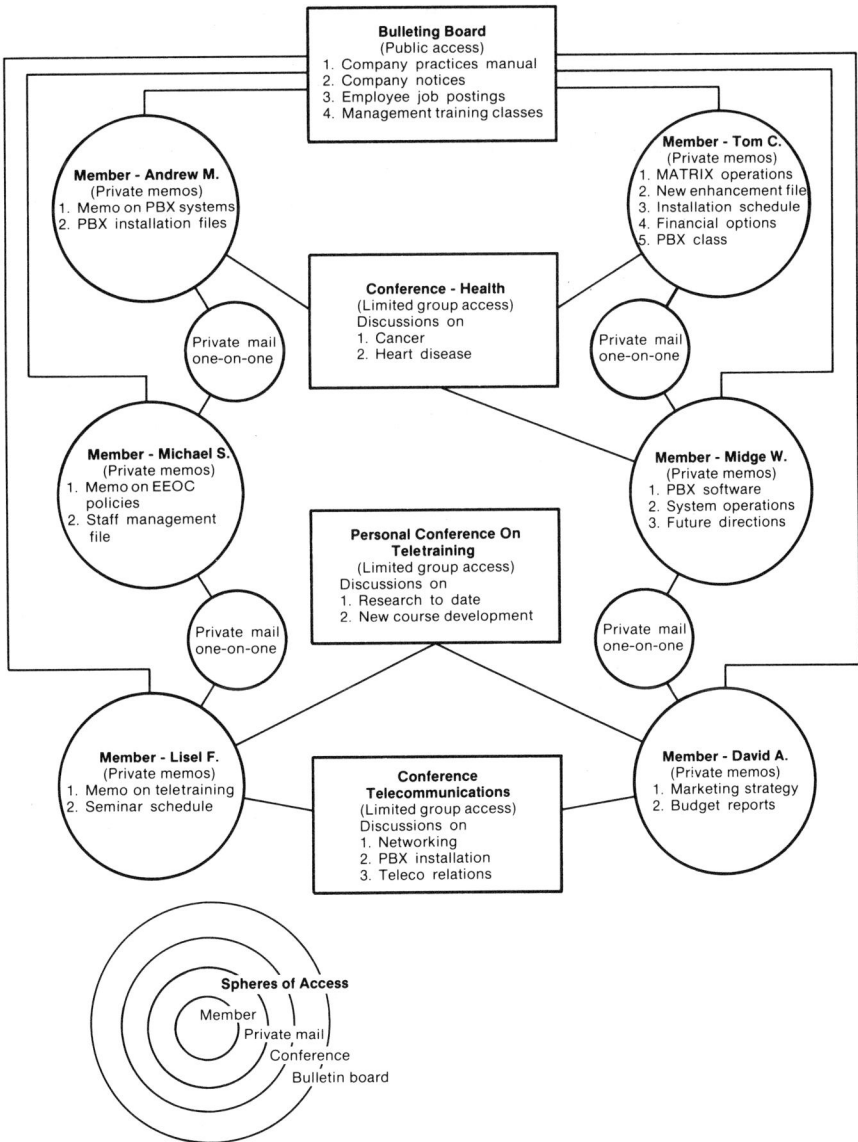

Bulleting Board
(Public access)
1. Company practices manual
2. Company notices
3. Employee job postings
4. Management training classes

Member - Andrew M.
(Private memos)
1. Memo on PBX systems
2. PBX installation files

Member - Tom C.
(Private memos)
1. MATRIX operations
2. New enhancement file
3. Installation schedule
4. Financial options
5. PBX class

Conference - Health
(Limited group access)
Discussions on
1. Cancer
2. Heart disease

Private mail
one-on-one

Private mail
one-on-one

Member - Michael S.
(Private memos)
1. Memo on EEOC policies
2. Staff management file

Member - Midge W.
(Private memos)
1. PBX software
2. System operations
3. Future directions

Personal Conference On Teletraining
(Limited group access)
Discussions on
1. Research to date
2. New course development

Private mail
one-on-one

Private mail
one-on-one

Member - Lisel F.
(Private memos)
1. Memo on teletraining
2. Seminar schedule

Conference Telecommunications
(Limited group access)
Discussions on
1. Networking
2. PBX installation
3. Teleco relations

Member - David A.
(Private memos)
1. Marketing strategy
2. Budget reports

Spheres of Access
Member
Private mail
Conference
Bulletin board

SOURCE: Courtesy Cross Information Company.

ages, one finds a large number of features but also limitations on BBS or electronic mail. Some of the basic considerations are:

- Speed—300 to 9,600 bits per second or more.
- Protocol conversion—will one system talk or send files to another and handle standard modem languages?
- Password protection—will privacy protection prevent one person from reading another person's mail? Are there individual password protection levels available? If one level of password protection is good, additional levels add greater security for privacy.
- Additional security measures—does the system offer the means of protecting other types of communication such as conferencing, the BBS, or private notepads?
- Micro-mainframe communications—are the files created by the BBS in a form that another PC, minicomputer, or mainframe system can understand? Is a special data converter required for communications to take place? Many software programs offer teletypewriter (TTY), American National Standards Institute (ANSI)—American Code for Information Interchange (ASCII—7 level), extended binary coded decimal interchange code (EBCDIC—8 level), or Baudot (teleprinter—Telex—5 level).
- Additional features—graphics, video pictures, and other images can be merged with words or data for communications. Conferencing, spreadsheet, and database management functions are becoming commonplace and interconnected with mail programs.
- Number of users—many systems allow an unlimited number of users (limited only by the capacity of the disk). Others limit the users to groups or committees.
- Disk and memory requirements—it is necessary to have sufficient memory to process or "bag" the mail that is sent to a BBS and take care of other housekeeping functions such as "talking" to the PC operating system, modem handshaking, and filing. The amount of disk required for a BBS can be extensive, especially when users are allowed to keep all their mail or send copies to all other members (such as a committee report), or when old BBS notes are not purged. (A list of some BBS, Email, and PC-based conferencing systems can be found in the Resource Guide.)

BBS used by telecommuters has been referred to as a "teleclub," "telecoffee," "telechat," or "electronic watercooler," and is usually considered to be a "place" to fraternize. For business purposes, however, the electronic mailbox within a BBS can be used as a drop-off point for work produced by the telecommuter.

Using computer, modem, and telephone lines, one can reach any number of public and private, free and subscriber electronic bulletin boards across the country. There are listings of bulletin boards available from most local PC clubs. Once you have dialed and reached the BBS, it is possible to post public messages, send electronic mail to other BBS users, and look through libraries of articles, data, and programs that are stored for members.

It is also possible for users to develop nationwide contacts among people with similar interests through using a BBS. The Independent Insurance Agents Association supports a free BBS where members can "swap business tips." H&R Block Company's CompuServe offers a work-at-home forum for telecommuters (costing $40 for start-up fee and $6 to $15/hour use), and bulletin boards are available for people with almost any type of interest, from the highly technical and scientific to the most mundane.

A Work-at-Home Special Interest Group (SIG) bulletin board has become popular with telecommuters who use the CompuServe network. Membership in a special interest group may be limited to a specific class of user or be open to any interested person. Users can peruse the message base contained in the SIG, then read and reply to messages of interest. Mining specialists networked via the "Miner's Underground" use their system to question peers on tricky problems.

Electronic mail (EM) features. There are many types of electronic mail, some of which are available for telecommuters. In the broadest generic sense, any system that sends a message or document in electronic form from one place to another can be considered electronic mail. See Figure 7–4. Each of the following concepts is a different type of electronic mail:

- Telex/TWX—see Figure 7–5.
- Postal electronic mail (e.g., ECOM).
- Facsimile.
- Point-to-point (often called station-to-station) communications (interoffice). See Figure 7–6.
- Intraoffice Computer-Based Mail System (CBMS).
- Message center—video message center with central printer and telephone dispatch.
- Broadband local area network—interfaced with microcomputers, a file server or facsimile machines.
- PBX-based electronic mail.

There may be other ways to categorize EM. In fact, some vendors consider "document distribution" to be a form of EM. Electronic mail is a system intended to move information electronically rather than manually, whether it is used to move documents, telephone messages, or corporate correspondence. Electronic mail users sometimes err in calling

FIGURE 7-4 Electronic Mail Touches on All Aspects of the
Communications Industry

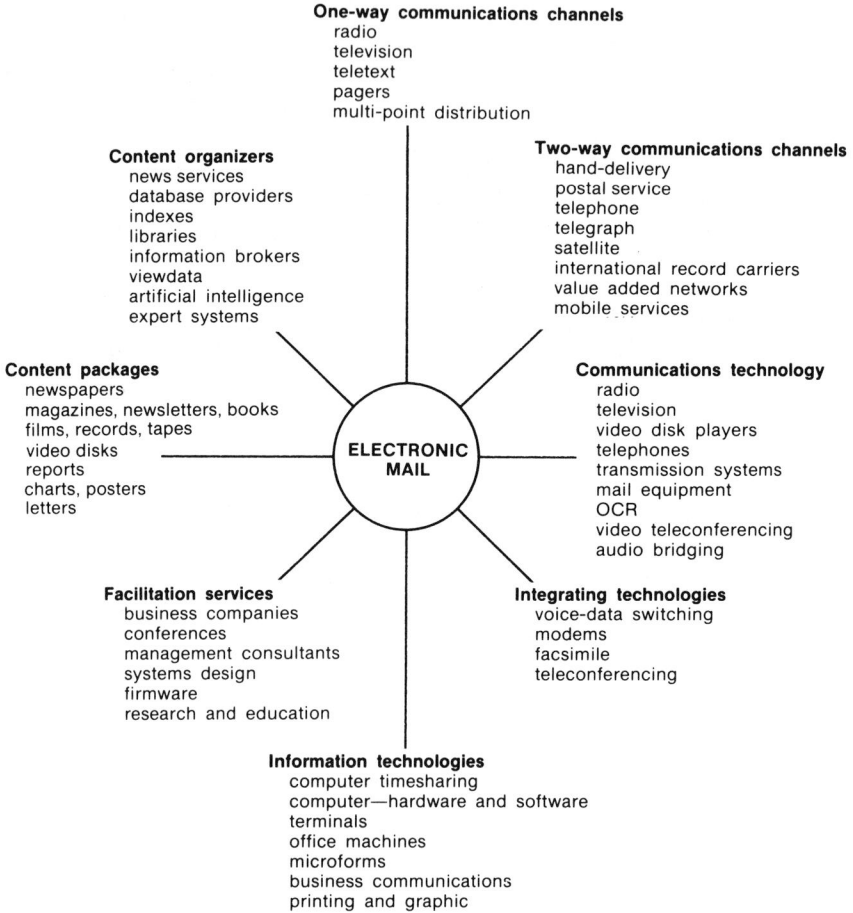

One-way communications channels
radio
television
teletext
pagers
multi-point distribution

Content organizers
news services
database providers
indexes
libraries
information brokers
viewdata
artificial intelligence
expert systems

Two-way communications channels
hand-delivery
postal service
telephone
telegraph
satellite
international record carriers
value added networks
mobile services

Content packages
newspapers
magazines, newsletters, books
films, records, tapes
video disks
reports
charts, posters
letters

ELECTRONIC MAIL

Communications technology
radio
television
video disk players
telephones
transmission systems
mail equipment
OCR
video teleconferencing
audio bridging

Facilitation services
business companies
conferences
management consultants
systems design
firmware
research and education

Integrating technologies
voice-data switching
modems
facsimile
teleconferencing

Information technologies
computer timesharing
computer—hardware and software
terminals
office machines
microforms
business communications
printing and graphic

FIGURE 7-5 Telex

Hard copy

Hard copy

International telex
service lines

Teletypewriter
(teleprinter)

Teletypewriter
(teleprinter)

FIGURE 7-6 Terminal-to-Terminal Electronic Mail

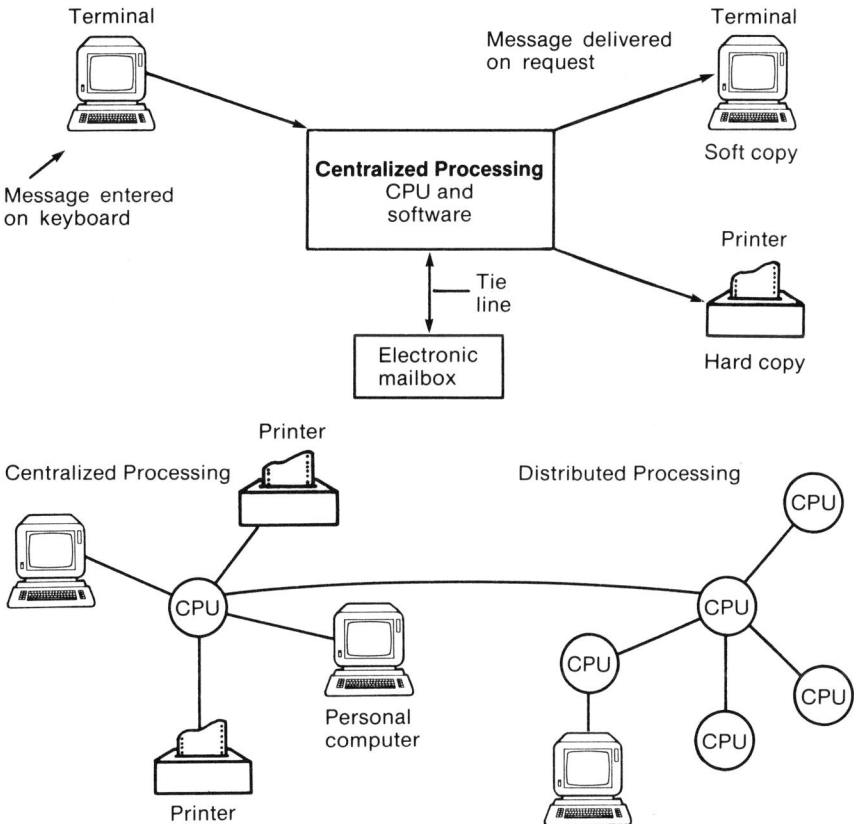

local area networks electronic mail networks, although EM is one of their primary capabilities.

This tutorial on electronic mail represents some but not all of the features found in electronic mail systems. For more information on electronic mail, read *Networking: An Electronic Mail Handbook,* by Thomas B. Cross and Marjorie Raizman (Scott, Foresman).

Today's electronic mail systems are rapidly becoming feature rich and feature diverse, integrating with spreadsheets, database management systems, thought processing, and other systems. Various EM systems are similar in certain respects but organized in different ways, just as Fords and Chevrolets are automobiles but are made with distinct characteristics and look different.

Electronic mail produced on word processors, for example, is usually transmitted to the recipient in real time. In contrast, other computers typically deposit mail in a central or host computer where it is kept until the recipient requests its delivery. PCs that rely on a central computer for electronic mail transmission can operate in both real-time and store-and-forward modes. Moreover, information or messages can be sent or broadcast to groups of people.

Unlike the U.S. Postal Service, the electronic mail system must first recognize the recipient before it creates or sends a message. Most EM systems also have certain protocols for creating a message. They require, for example, that one know how to access commands by using special-feature buttons which allow the user to review a message before sending it.

Some commonly found electronic mail commands for sending messages in a store-and-forward approach are:

- Send message—puts message in recipient's mailbox within the system.
- Forward message—allows a message to be read and forwarded to one or more readers.
- Quit—stops activity and allows the user to leave the mail system.
- Help—brings information and on-line instructions onto the screen.
- Timed message delivery—allows the writer to create a message for future delivery. This feature can be used as a timed reminder to the writer.
- Group—allows message distribution to a group of people.
- Copy—sends carbon or blind copies.
- Registered/forced reply—allows the sender to be informed when a message is received and, in some cases, forces the recipient to reply.

Certain systems provide the user with a menu of commands. See Figure 7–7. This is an easier approach than learning a complicated list of commands. However, users become quickly bored by moving through tedious menu listings and prefer short, abbreviated commands such as "SM" for send mail or "RM" for retrieve mail.

Typical features for reading or receiving mail are:

- Acknowledge or received—this lets the sender know that the message was received and allows a reply.
- Again—allows the message to be read again.
- Print—tells the computer to have the message printed out on a hard-copy printer.
- Hold—indicates that the recipient received the message but will not reply at this time. It also allows the recipient to hold the message until later.
- Save/File/Copy/Delete—puts the message in an electronic file cabinet or "wastebasket."

FIGURE 7-7 Specifications—Typical Computer Conference System

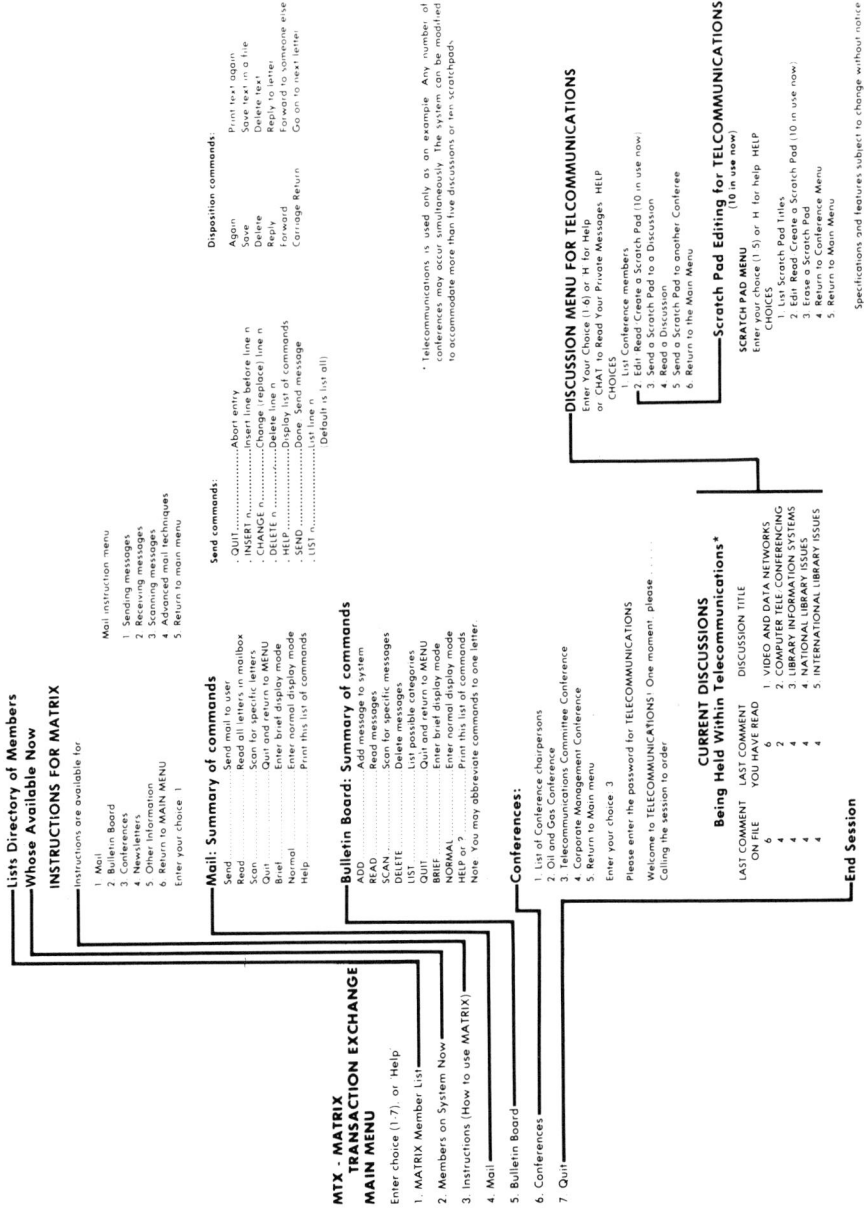

List Directory of Members

Whose Available Now

INSTRUCTIONS FOR MATRIX

Instructions are available for

1 Mail
2 Bulletin Board
3 Conferences
4 Newsletters
5 Other Information
6 Return to MAIN MENU
Enter your choice 1

Mail instruction menu

1 Sending messages
2 Receiving messages
3 Scanning messages
4 Advanced mail techniques
5 Return to main menu

Mail: Summary of commands

Send .... Send mail to user
Read .... Read all letters in mailbox
Scan .... Scan for specific letters
Quit .... Quit and return to MENU
Brief .... Enter brief display mode
Normal .... Enter normal display mode
Help .... Print this list of commands

Send commands:

QUIT .... Abort entry
INSERT n .... Insert line before line n
CHANGE n .... Change (replace) line n
DELETE n .... Delete line n
HELP .... Display list of commands
SEND .... Done Send message
LIST n .... List line n
                     Default is list all)

Disposition commands:

Again .... Print text again
Save .... Save text in a file
Delete .... Delete text
Reply .... Reply to letter
Forward .... Forward to someone else
Carriage Return ... Go on to next letter

Bulletin Board: Summary of commands

ADD .... Add message to system
READ .... Read messages
SCAN .... Scan for specific messages
DELETE .... Delete messages
LIST .... List possible categories
QUIT .... Quit and return to MENU
BRIEF .... Enter brief display mode
NORMAL .... Enter normal display mode
HELP or ? .... Print this list of commands
Note You may abbreviate commands to one letter.

Conferences:

1 List of Conference chairpersons
2 Oil and Gas Conference
3 Telecommunications Committee Conference
4 Corporate Management Conference
5 Return to Main menu

Enter your choice 3

Please enter the password for TELECOMMUNICATIONS

Welcome to TELECOMMUNICATIONS! One moment please . . . . . .
Calling the session to order

CURRENT DISCUSSIONS
Being Held Within Telecommunications*

LAST COMMENT ON FILE	LAST COMMENT YOU HAVE READ	DISCUSSION TITLE
6	6	1 VIDEO AND DATA NETWORKS
4	2	2 COMPUTER TELE CONFERENCING
4	4	3 LIBRARY INFORMATION SYSTEMS
4	4	4 NATIONAL LIBRARY ISSUES
4	4	5 INTERNATIONAL LIBRARY ISSUES

* Telecommunications is used only as an example. Any number of
conferences may occur simultaneously. The system can be modified
to accommodate more than five discussions or ten scratchpads.

DISCUSSION MENU FOR TELECOMMUNICATIONS

Enter Your Choice (1-6) or H for Help
or CHAT to Read Your Private Messages: HELP
CHOICES
1 List Conference members
2 Edit Read Create a Scratch Pad (10 in use now)
3 Send a Scratch Pad to a Discussion
4 Read a Discussion
5 Send a Scratch Pad to another Conferee
6 Return to the Main Menu

Scratch Pad Editing for TELECOMMUNICATIONS
(10 in use now)

SCRATCH PAD MENU
Enter your choice (1-5) or H for help HELP
CHOICES
1 List Scratch Pad Titles
2 Edit Read Create a Scratch Pad (10 in use now)
3 Erase a Scratch Pad
4 Return to Conference Menu
5 Return to Main Menu

Specifications and features subject to change without notice.

MTX - MATRIX TRANSACTION EXCHANGE MAIN MENU

Enter choice (1-7), or Help

1. MATRIX Member List
2. Members on System Now
3. Instructions (How to use MATRIX)
4. Mail
5. Bulletin Board
6. Conferences
7. Quit

End Session

SOURCE: Copyright Cross Information Company.

- Append—allows a note to be attached to the bottom of a previously written message.
- Reply—invokes the text editor so that a reply can be created.
- Forward—routes the message to a designated person for consideration. A message can be appended with comments, then sent on to its final destination.

There are a myriad of other features found on electronic mail systems. They allow creation of multilevel passwords, listings of messages sent or received, verification of messages, priority position, and other features. In Cross/Point™ developed by Cross Information Company, users have access to "windows" which view up to eight different pieces of mail, BBS notes, and conference comments. They also permit the user to make notes at the same time.

Software standards like the CCITT X.400 will eventually allow different electronic mail systems to interact. This will permit telecommuters to link up without worrying about technical issues.

Features of notepads or private work areas. Each writer has private, secure (optional password protected) text files or memo areas which are kept on-line. These electronic files function as an electronic desk, with electronic file folders that contain memos, plans, and correspondence. Any of this material can be sent to members.

From the telecommuter's perspective, computer or text teleconferencing allows documents to be shared more easily, information exchanged on a group basis, and meetings to go on as long as necessary.

Computer teleconferencing systems provide the following additional benefits as well:

- *Outlining*—A computer is ideally suited for generating and organizing an outline of unlimited length. It can be used to add, delete, modify, and reorganize topics in a way that most paper-based outlines cannot. In addition, separate outlines or chapters can be linked at any point in either a grid or tree-trunk form. These relationships can be displayed on the computer terminal.

    The command to "prioritize" will organize a list of topics by their importance. The "randomize" command puts a list of topics in random order for possible associations. Grouping concepts by their common attributes permits general observations. The "categorize" command creates an outline structure from grouped topics on a list, and other commands provide additional links to organize, identify, flag, and retrieve information.

- *Charting*—When using a chalkboard, one often wants to save what has been written but needs the space for other material. Computer teleconferencing systems are ideally suited for saving work while new text is being created. In addition, the networking capability allows text to be sent to other people for review, comment, and correction. See Figure 7–8.

FIGURE 7-8   Futuristic Thought-Processing System

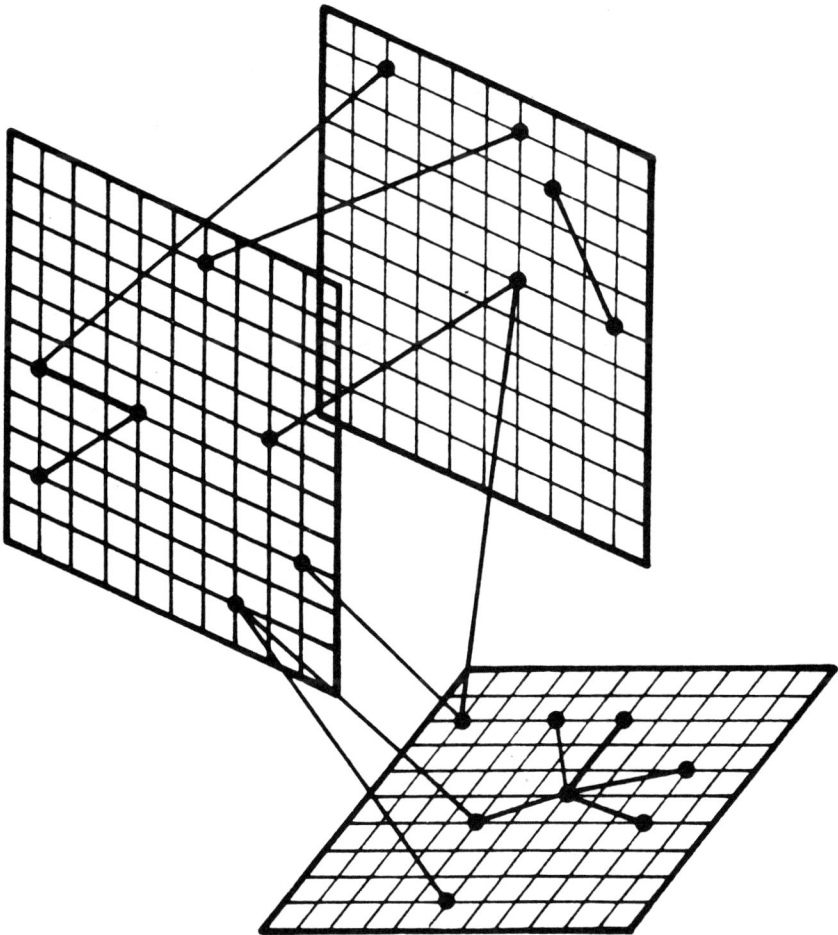

SOURCE: Courtesy Cross Information Company.

- *Networking*—Networks allow people to coordinate ideas, manage, reorganize, and present them for review. The system permits easy filing, searching, and editing. Creating ideas, linking them through multidimensional outlines, and charting their relative positions becomes a far more exciting and creative process than handling a traditional filing system.
- *Idea exchanges*—An "idea exchange point" is an organized information outline the telecommuter can use for group networking. Each idea exchange can be divided in various ways, into projects, chapters within a book, lectures within a course, or issues within a discussion. Idea points allow a group to share information in a central

idea exchange area or permit many-to-many or group networking. Work areas are accessed much like normal meetings, except there can be an unlimited number of idea exchanges (main topics) with a corresponding unlimited number of idea points (subtopics).

- *Status and tracking*—This function provides the conference member with information about new ideas that have been entered into the conference files, new mail, and information concerning project activities.
- *Management reports and directories*—These files inform network members how long an individual has been working at an activity. This is useful information for workers at distant locations, for people using charge-back systems, and for management reports. The directory is a useful system for finding out which idea exchanges and what members are using the system, or who is interested in a specific issue.
- *Searching*—Text searching of outlines, networks, membership, discussions, and personal notepad areas can save time and lead exchange members to far better organization. Furthermore, after a search is completed, the results can be instantaneously sent throughout the system, reorganized into a report, or filed.
- *Gathering*—Once information has been amassed and placed in separate idea files, it can be reorganized into a summary report via the "gathering function." This function collects personal memos, mail, and ideas, and organizes them in any way the telecommuter desires.
- *Real-time*—Although most computer teleconferencing activities are performed in nonreal time, this feature provides an on-line conversation mode. Because the system "knows" who is presently on line, it can help organize a "conversation" session (like a telephone conference or face-to-face meeting), except that statements are made in text form. In addition, a verbatim transcript of the event is made automatically.

One of the more exciting software systems that can be used is called cc:Share™. It works with any standard program such as a spreadsheet, word processing, or database management system. With cc:Share™, two users can simultaneously work on the same program in real time. In effect, they can look over each others' shoulders although they are a thousand miles apart.

- *File handling*—This operation is in increasing demand. Most systems can read ASCII files but have limitations on the amount of text they can transfer. Some systems allow the uploading and downloading of text of ASCII files but cannot read them as they were written. Printing options include spacing, selected files, outlines, charts, margins, headings, annotations, numbering, and status reports.

- *Other features*—The joining feature merges text by lines, words, or sections. Dividing reduces text into lines, words, or sections. Sorting performs a variety of activities by arranging text in ascending or descending order, by charts, or other options.

  Inserting and deleting allows input of new text within existing or new areas. Most computer teleconferencing systems have commands for text manipulation. Editing functions offer a wide range of options, including interfacing word processing software and internal editing commands.

- *Other program options*—One of the most exciting developments in screen display software is having different files displayed simultaneously on the screen. This feature should facilitate remote work. The screen is divided into small display areas called "windows." These can present text, graphics, notepads, and in the near future, video frames. Window displays allow the user to read comments from the conference discussion in one window, make notes in another, search a database in a third window, and receive incoming mail in a fourth window.

  In this way, the telecommuter could, and probably will, be connected to the organization's computer system for long periods of time. Because windows will allow the user to send and receive information at the same time, he or she can be more efficient and productive.

- *Security features*—Computer teleconference systems, like gatekeepers, recognize certain user accounts and provide them with security and privacy. In an open computer teleconference, all members may read all items and contribute at will. Normally, participants are not permitted to change files that other people created, although it is frequently useful to give at least one member editorial power to add or delete text. The editorial function is determined by the system designer.

A computer teleconferencing system can have many ongoing teleconferences within. They are often called conferences and committee or discussion subgroups. The software tracks conference participants and logs who has seen which items within each conference. A telecommuter or user who logs on is notified when new material is present.

Computer teleconferencing systems make the learning process much easier by providing menus, electronic road maps to the logical network, and on-line help to guide inexperienced people through the process. The following aspects of computer software define the conference structure:

- Specified roles for participants that permit or restrain their access to information, ability to vote on issues, etc.

- Selective communication among conferees for polling on issues, analyzing, and giving feedback of results.
- Participation by either a real name, an assumed name, or anonymously. This produces a "task-oriented," rather than "status-oriented," meeting.
- Capability to access, file, cross reference, and retrieve information from the system.

Audio Teleconferencing

An audio teleconference is a real-time meeting that can be established by dialing in to a "meet me" bridge, or dialing out from an audio bridge or telephone company. For more information on audio bridging features, see Appendix A. In the dial-out process provided by a number of telephone companies, a conference operator or controller calls all participants sequentially prior to the meeting, then bridges them together. A dial-out meeting can take up to 30 minutes to set up because people may be away from their desks or talking on their telephones. The telecommuter can participate in an audio teleconference from any available telephone.

When using the dial-in or "meet me" bridge, often provided by a service bureau or a long-distance telephone carrier, individual teleconferees call into the bridge at a specified time. Using a "meet me" bridge reduces audio teleconference setup time by eliminating such major delays as occur when one can't take calls or there are poor connections. Companies providing these services are listed in the Resource Guide.

An audio bridge is a device that generally allows up to 48 people (or more, depending on the system used) to be connected simultaneously. Telephone lines go into the bridge where sound is amplified. Bridge amplification modifies participants' voice quality, and noise is filtered out. An operator controls the bridge console and arranges a predetermined time for the teleconference.

Audio bridges are now being automated. This permits conferees to hold daily multipoint teleconferences without the need for a conference operator to participate. In some systems, a microprocessor coupled with a voice synthesizer greets the teleconference participants and instructs them to introduce themselves to the conferees. People access the system via tones generated by a touch dial pad on the user's telephone.

The following are some of the more viable audio teleconferencing applications:

- Coordinating internal administrative affairs.
- Interfacing with home and field offices.
- Developing product releases and telemarketing.
- Field personnel training.

Operator Setting Up an Audio Teleconference

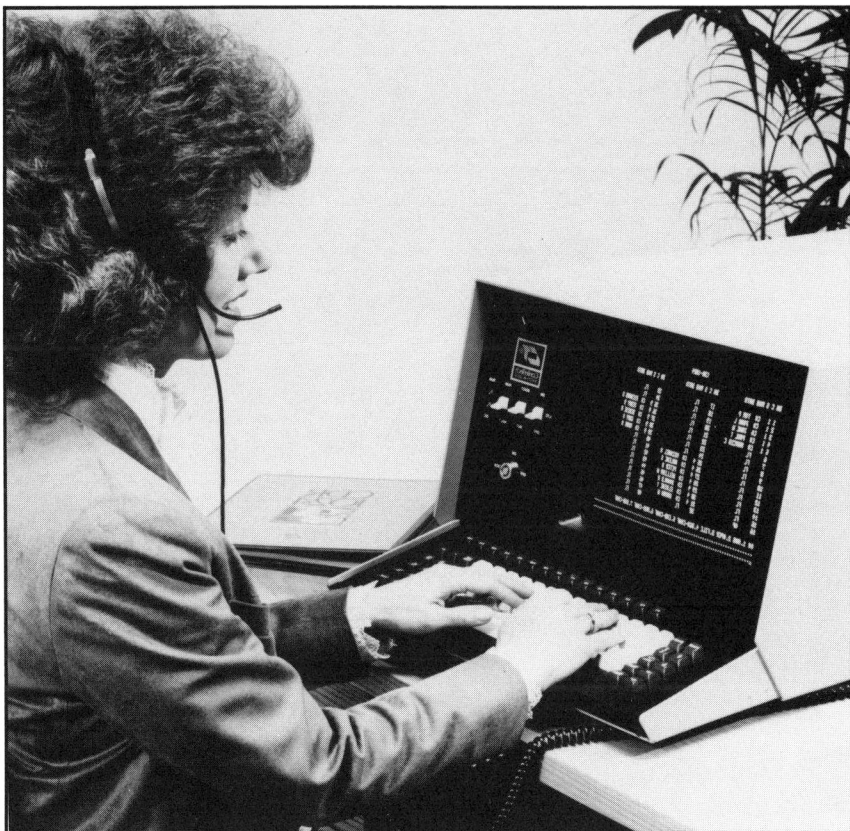

Courtesy Confertech International, Inc.

- Coordinating remote manufacturing/operations.
- Public relations and product development.

For the use of audio bridging and teletraining systems, see Appendix B.

Audio-Graphic Teleconferencing

A conference is known as an audio-graphic teleconference when visual transmission equipment supplements audio transmission devices. Transmission of graphics reinforces the audio information transmitted and enables design, engineering, editorial, and other collaborations to occur in real time. Graphics can also be sent ahead of or after meetings.

FIGURE 7-9 Instructor Using an Electronic Black Board to Teach

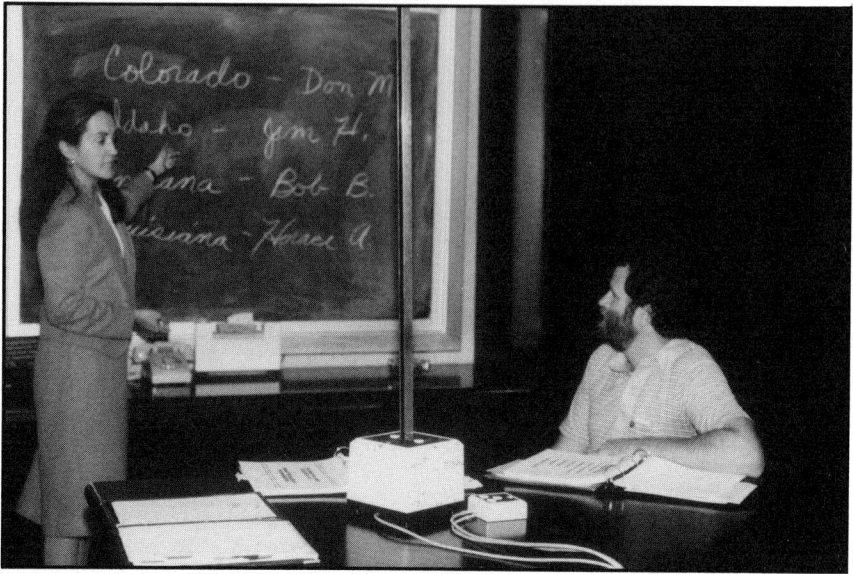

Courtesy Virginia A. Ostendorf, Inc.

Various audio-graphic teleconferencing equipment includes:

- Telewriters.
- Multicolored electronic pens.
- Telewriter video terminals.
- Integrated workstation and speakerphone.
- Electronic chalk and marker boards—see Figure 7–9.
- Personal computers.
- Lapboards for freehand drawing.
- Mouse input devices.
- Facsimile-like (scanning) devices.

The addition of graphics to audio teleconferencing enhances specialized telecommuting applications such as discussions of:

- Engineering designs.
- Building layouts.
- Advertising.
- Interactive equations.
- Legal briefings.

Visual-Graphic Teleconferencing

Visual graphics and freeze-frame video are extensions of the audio-graphics concept in teleconferencing. Freeze-frame video adds a visual dimension, and at an added cost, supports the audio element. Synchronization of media merges their separate advantages and provides an effective form of teleconferencing for a fraction of the cost of full-motion video teleconferencing.

The advantages of visual-graphic teleconferencing are:

- The emphasis is on moving graphics and data rather than on people.
- Spreadsheet software, featuring columns, charts, graphs, and text allow for system application in many areas.
- Software programming on a personal computer (PC) is a substantially faster, simpler, and cheaper process than full-motion video teleconferencing. In addition, editing involves only inserting a floppy disk into the PC drive and executing desired changes.
- This type of system uses the large number of computer mainframes and PCs that many companies have. By stressing computer graphics over video, these companies can use their existing data processing resources to make video teleconferencing cost effective. This breakthrough will probably make the personal computer the dominant and creative force in teleconferencing.

Video Teleconferencing

Full-motion video teleconferencing enables people at two or more locations to have contact almost as though they were seated in the same room. Applications for this system in telecommuting situations may be most effective when information is sent from a downtown main office to a neighborhood work center. For example, Merrill Lynch is installing full-motion links from Manhattan to a New Jersey training center about 50 miles away. These allow people to telecommute instead of driving into the city.

The four broad categories of video teleconferencing are:

1. Ad hoc—Using broadcast-type facilities located in television stations and hotels. The goal is to reach thousands of people for a onetime exposure. The cost is between $14 and $200 per person per meeting. See Figure 7–10.
2. Interstate corporate—Using private or public rooms located on the customer's premise or in a metro area. This is typically used for 6 to 10 people per location and costs $2 to $8,000 per meeting, depending on the duration and distances.

FIGURE 7-10   Satellite Teleconferencing

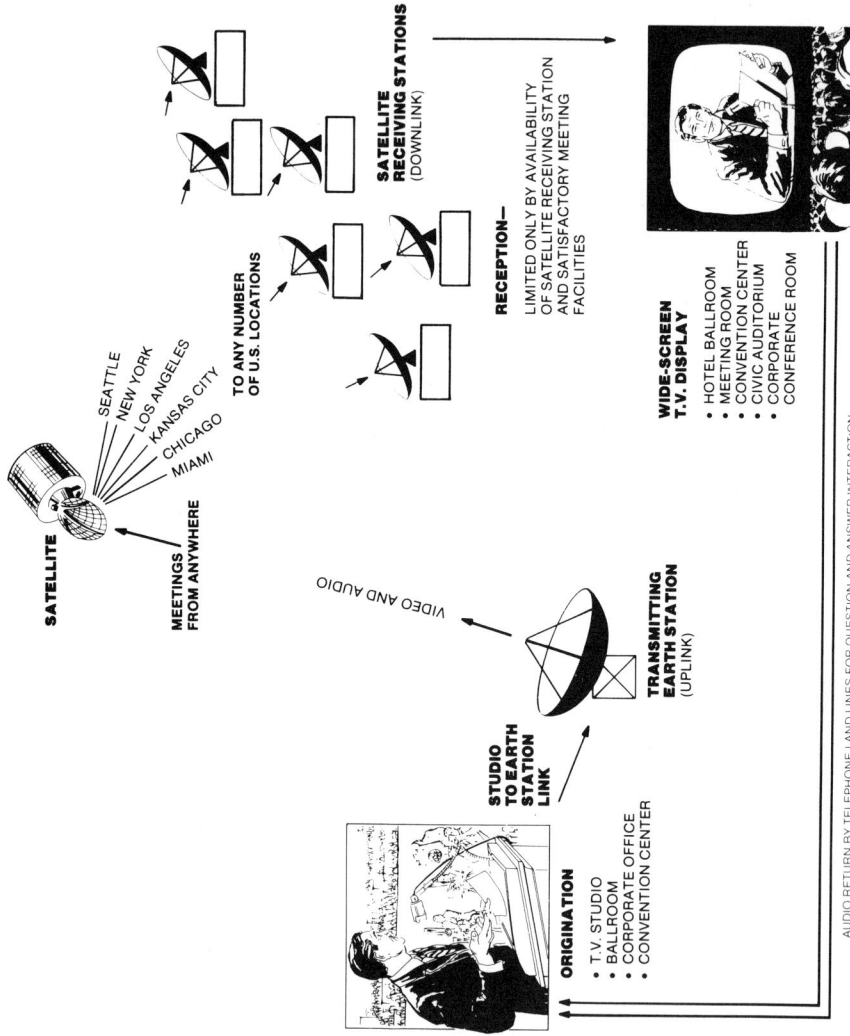

SATELLITE

MEETINGS
FROM ANYWHERE

SEATTLE
NEW YORK
LOS ANGELES
KANSAS CITY
CHICAGO
MIAMI

TO ANY NUMBER
OF U.S. LOCATIONS

SATELLITE
RECEIVING STATIONS
(DOWNLINK)

RECEPTION—
LIMITED ONLY BY AVAILABILITY
OF SATELLITE RECEIVING STATION
AND SATISFACTORY MEETING
FACILITIES

WIDE-SCREEN
T.V. DISPLAY
• HOTEL BALLROOM
• MEETING ROOM
• CONVENTION CENTER
• CIVIC AUDITORIUM
• CORPORATE
• CONFERENCE ROOM

VIDEO AND AUDIO

STUDIO
TO EARTH
STATION
LINK

TRANSMITTING
EARTH STATION
(UPLINK)

ORIGINATION
• T.V. STUDIO
• BALLROOM
• CORPORATE OFFICE
• CONVENTION CENTER

AUDIO RETURN BY TELEPHONE LAND LINES FOR QUESTION AND ANSWER INTERACTION

Videostar Tele-meeting® Network.

SOURCE: Courtesy Videostar.

3. Intrastate local/campus—A set of privately owned rooms used for daily business meetings or education within a corporate campus or local neighborhood work center.
4. International—Incorporates both ad hoc and interstate conferencing. The corporate market value of this category is perceived to be high.

Some of the key features of video teleconferencing are:

- Real-time delivery of video images.
- Allows for perception of personal presence.
- Participants' body language and emotions can be seen.
- Allows for rapid decision making.
- Ideal for group meetings as opposed to one-to-one discussions.
- Allows "crisis meetings" to take place quickly regardless of conferees' locations.
- Allows "high-impact" Hollywood-style events.
- Often requires large capital commitment.
- Requires costly ongoing overhead and maintenance.

Video teleconferencing problems include the high cost of such systems that is due to their degree of complexity and to the small number of systems suppliers. Most suppliers provide single components, such as audio speakerphones, television cameras, transmission systems, and so forth. Only a few companies are emerging as total systems suppliers. See Figure 7–11. They use approaches that reduce the corporation's video teleconferencing costs. Although video teleconferencing is a valuable business tool, it can be prohibitively expensive if acquired for individual telecommuters.

Still-Frame Teleconferencing

Still-frame, freeze-frame, or slow-scan video teleconferencing is currently, and may be in the future, the easiest and most cost-effective means for communicating visually in telecommuting environments. High-quality slow-scan video is much like a slide show. "Still-motion" images are sent every 8 to 35 seconds, depending on the bandwidth of the circuit (e.g., normal telephone lines average 35 seconds per frame). In most slow-scan systems, images are built upon the screen from left to right or top to bottom.

The cost of slow-scan systems is significantly lower than that of full-motion video. And while full-motion video provides a very natural environment for conducting meetings, still images are of particular use for person-to-person communications where action is neither important nor desired. This is true, for example, in transmitting a printed page, blueprint, memo, or photography. It is also true in transmitting a typical

FIGURE 7-11 Typical Video Teleconference Room Layout

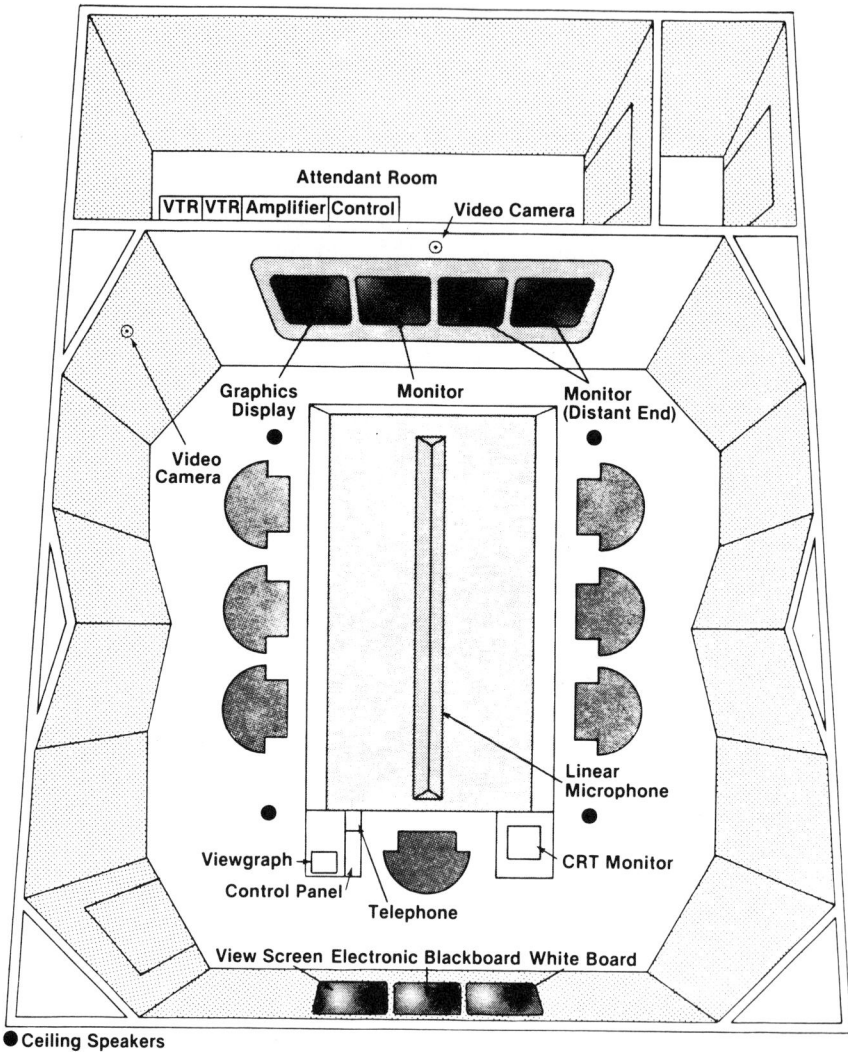

SOURCE: Copyright 1984 Cross Information Company.

meeting or in a classroom situation where the speaker shows slides, view-graphs, flip charts, chalk drawings, or magazine excerpts.

Among the features of slow-scan teleconferencing are:

- It uses normal telephone lines rather than broadband circuits. This is important for telecommuting because it may be some years before very high speed (1.544 megabits per second) communications lines are available for residential areas.

Slow-Scan Video Teleconference

Courtesy Colorado Video.

- It costs very little compared to full-motion video. Slow-scan systems are also physically small. Slow-scan video converters or "video modems" are small enough to be located on a desk.
- It helps solve problems faster by improving access to people and information.
- It provides portability and usability at remote sites. In telecommuting, the need for "briefcase" portability is desirable, as the worker may not be at home or in any one place all the time.
- It allows pictures or presentations to be recorded on an audiocassette recorder or computer disk for later use. Thus, it can be used to present case studies demonstrating techniques, etc.
- It provides flexible time and distance options. Slow-scan can be used in a store-forward (asynchronous or nonreal-time) situation where pictures, documents, or training sessions can be delivered to telecommuters, even when they are not at home. This has also been called "up loading" or "down loading," depending on whether the information is being sent to or from the office.
- It makes available high- and low-resolution video systems. The resolution is of sufficient quality that radiologists can serve many cities at distant locations.
- Provides nonmoving pictures for presentations similar to a 35mm slide show. There is comparatively little need for motion since

most corporate presentations to executives, during training sessions, or for typical office communications are given by slides or overhead projector foils.

In slow-scan teleconferencing, single-picture transmission provides enormous savings in channel capacity when compared to conventional video use. The current trend is to use conventional video cameras, monitors, and other system components in conjunction with "scan conversion" devices. These reduce the bandwidth of the television camera's output for transmission over voice grade or normal telephone lines. This ratio is achieved by stretching signal time from 30 pictures per second to 1 picture per 7 to 35 + seconds and/or by reducing visual resolution. Because telephone line bandwidth is expensive, this represents a large savings in transmission costs.

IBM's slow-scan video teleconference system presently comprises over 110 rooms worldwide. Using the IBM system, it is possible to transmit voice and black-and-white or color video images of people and graphics as well as to send high-speed facsimile and electronic storage of documents for recall and later transmission. Security arrangements allow for presentation and discussion of confidential IBM information.

Slow-scan video has a number of telecommuting applications. Some of these include:

- Teleteaching—company or university courses.
- Telemeetings—engineering design meetings.
- Telepublishing—news and information publishing.
- Telemedicine—remote diagnostics, including X-ray analysis.
- Telemonitoring—environmental monitoring and security.

Voice Mail

Voice mail (VM) is the counterpart of electronic mail. Each system has advantages, yet both are communications technologies that represent one of the most important considerations in telecommuting today. See Figure 7–12.

Voice mail is another technology that extends and enhances communications and can be helpful to the telecommuter. Voice mail or voice store-and-forward systems save and digitize analog speech on a hard disk for later retrieval. Stored on a computer disk system, a message can be processed in the same way as other forms of electronic communications. Typically, voice mail users are assigned "voice mailboxes" to which digitized voice messages can be delivered for later retrieval.

Voice mail systems improve productivity and control telephone costs. Some features found in these systems enable users to:

- Record-and-store messages.
- Stop, start, listen, and slow, normal, fast controls.

FIGURE 7-12   Voice Mail

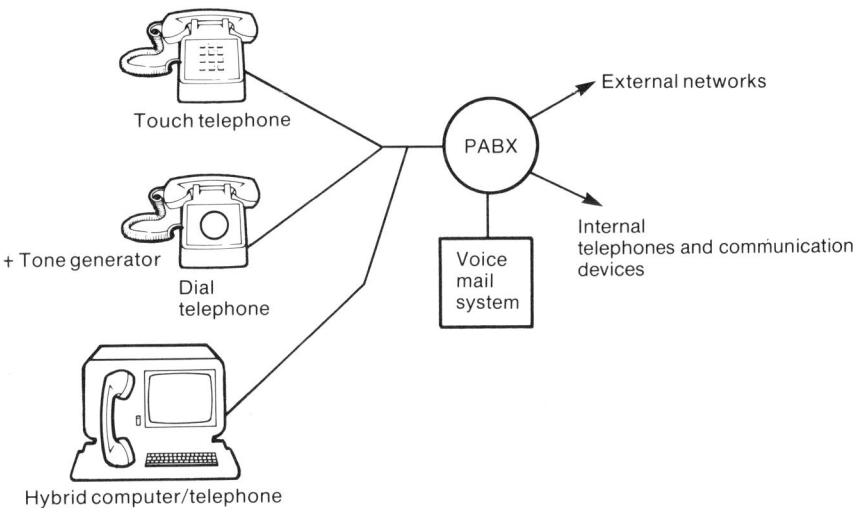

- Retrieve messages sequentially.
- Retrieve messages by random selection.
- Forward messages to individuals or groups.
- Distribute to lists of people stored in the system.
- Classify messages, such as urgent and timed-delivery.
- Customize information, such as traveling schedules.
- Secure accessibility to confidential information.
- Compress pauses for more concise messages.
- Have prompt and help functions.

See Appendix C for voice mail feature comparison.

Voice mail does much more than telephone message recording devices by processing as well as communicating messages. This enhances the value of information sent throughout the organization, helping organizations to:

- Control time spent recording and listening to messages.
- Identify and retrieve messages by individual's name, time, and date sent.
- Classify messages—as normal, quick-ring for immediate delivery during business hours, urgent for delivery after business hours, or for timed delivery.
- Categorize messages—as new (not yet listened to), pending (awaiting action or reply), old (listened to and being stored in the system), and outbound (by sender).

- Provide quality voice reproduction, including retaining a speaker's expression and inflections.
- Achieve a continuous, natural flow of words throughout recorded messages by elimination of pauses.
- Secure information with passwords. For example, one's self-created password allows for listening to all information. A secretary's password might be used to ascertain who sent messages. A guest password could be given to vendors or family members and friends.
- Schedule and distribute messages by specified date and/or time.
- Reduce long-distance telephone costs by transmitting messages during lower-rate hours.
- Decrease employee hours spent on the telephone by reducing or eliminating the need to call back, or for secretaries to transcribe messages.
- Record a variety of information in addition to messages, such as progress reports on particular tasks or projects, or of travel schedules. These are quickly accessed by secretaries and easily updated by principals.
- Send and receive recorded messages.
- Store messages for prompt retrieval, or keep them for longer periods.
- Send messages to individuals or groups.
- Transmit messages to people on distribution lists stored in the system.
- Append messages with comments for rerouting to others.
- Confirm receipt of messages sent.

The advantages of voice mail are:

- Communications sound natural.
- Recipients can verify sender by voice recognition.
- Message creation is rapid and efficient.
- Telephones are readily available.
- Rapid response to important messages is possible.
- Messages include emotional intent.

The disadvantages of voice mail are that it:

- Requires recipient to take notes.
- Is difficult to edit and annotate.
- Is difficult to document for record-keeping.
- Can be hard to send lengthy communications.
- Loses emphasis when transcribed in written form.
- Requires expensive system overhead in transmission time and computer management.

Cost justification for using voice mail systems is based on the time one saves by eliminting repeated calling. In at least three out of four instances, telephone calls do not reach the intended party, and "telephone tag" ensues. The time that is saved can be estimated and the cost per hour computed to determine the projected "hard" dollar savings. Since the estimated savings are also in "soft" dollars (improved communication and better decision making), they are difficult to identify and quantify.

Radio Teleconferencing

Because radio teleconferencing applications for telecommuting are limited, this subject will not be discussed beyond stating the following key characteristics of the system:

- It is an excellent transmission medium for use between widely dispersed locations and where the telecommuter may not be reached by most forms of transportation, e.g., Alaska.
- It can be connected to the telephone network.
- It is low cost compared to other communications systems.
- It is not often used when security is required.
- It requires special high-frequency receivers.
- It can be used as a paging system.

Cellular Telephone

Cellular radio systems represent the next step in the development of mobile telephone systems. The key advantage of cellular radio is its ability to provide for possibly thousands of mobile telephones in one area. This has already spurred the development of the "office without walls" approach to business that allows people to take calls from almost anywhere. Some cellular systems offer voice mail which enables the caller to also receive information at anytime. Nearly all major cities across the United States have operational cellular systems.

The concept of cellular radio comes from the structure of cells, much like those found in a beehive. Within each cell, there is a transmitter and receiver. Unlike the older mobile telephone systems which covered areas of 100 square miles or more, modern cells are approximately 10 miles in diameter. These smaller cells offer better mobile telephone reception as well as an increased number of channels. If an automobile equipped with a cellular telephone moves from one cell (area) to another, the systems automatically provide another channel.

Cellular radio systems may be used by telecommuters who find themselves on the road, between offices, and going to and from home. There is also the suggestion that cellular radio systems might be used inside the office building in place of wiring. A cellular switchboard or PBX system would connect telephones (either desk or portable) as well as office machines, including PCs, data terminals, modems, facsimile, and other devices, without being involved in wiring problems that exist today in most office buildings.

Security and privacy issues are certainly a question in using radio frequency (RF) devices. However, tapping into wires is easily performed as well. Infrared systems may offer the same "wireless" flexibility without the radio frequency issues such as ease of listening in, frequency congestion, and FCC licensing. Moreover, the cellular PBX radio system is probably limited to within a building or even an office floor.

Cellular radio systems can be used with modems to send data communications from PCs to remote databases at the office or elsewhere. However, simpler "rovaphones" that are available for less than $50 can be used in data communications applications when range is limited to the back porch or pool.

Cellular radio systems will play an important part in both telecommuting settings and in the development of advanced communications networks. This suggests that the automobile or other mobile facilities (boats, planes, trains) will be increasingly important in both where and how business is conducted.

BARRIERS TO ELECTRONIC COMMUNICATIONS

There are a multitude of ingrained office meeting habits which inhibit or restrict interpersonal communications and affect the successful integration of a telecommuting program. These habits interfere with the effective use of electronic communications systems. Telecommuters should understand that electronic communications do not replace local or long-distance travel but merely offer a supplement for many face-to-face meetings or presentations. Instead of driving to every company meeting, telecommuters should sometimes forgo office "perks" and corporate visibility and transmit their information via electronic systems.

Telecommuters should also recognize that meeting electronically offers them an alternative type of communications and possibly more impact. In speaking, one must often say words twice before they are really understood. In text systems, this redundancy can be omitted, or it may reduce the value or impact of a message. In fact, written text can have a more powerful meaning to the recipient than the sender intends. In telecommuting, this is an important consideration.

Telecommuting, like any other new technology, is faced with strong resistance. There are still many people who hope that interest in office automation, computers, and telecommuting will go away so that they can get back in the car or plane and conduct business in the way they have for decades. In order to be prepared for this resistance, telecommuting planners must be well prepared to present the case for "hard" dollar savings as well as for personal employee "soft" dollar benefits.

Some of the problems associated with most forms of teleconferencing (for a telecommuting program) include management's:

- Unwillingness to learn new procedures.
- Fear of improper equipment operation and connection.
- Fear of failure when teleconferencing.
- Unwillingness to prepare organized material.

To proceed with implementing this type of system, it is usually necessary to develop a "business case" and examine all teleconferencing technologies to understand how they work together in solving management problems, whether for telecommuting for other business activities. (Refer to the Resource Guide.) For those who have used electronic communications in telecommuting and other applications, life has been transformed. These tools are fast and efficient, and often more reliable than other forms of communication, including face-to-face meetings.

Conclusion

When managers are asked to list their biggest problems, after mentioning the problem of costs, they generally complain about the lack of communication in their operations. Electronic systems—whether teleconferencing, voice mail, text mail, or others—is neither a panacea for, nor a cure for poor human communication, whether it takes place in the office or during telecommuting. "If you are boring during a face-to-face meeting, you will still be boring in a teleconference," teleconference participants are frequently heard to say.

At the same time, where electronic systems supplement—not replace—face-to-face activities, there has been a marked improvement in overall communications, and they have been a driving force in enhancing the ability of employees to work independently of location. In essence, technology can improve communication but cannot create quality where it does not exist.

NOTES

1. William M. Bulkeley, "Better than a Smile: Salespeople Begin to Use Computers on the Job," *The Wall Street Journal*, September 13, 1985, p. 25.

2. William M. Bulkeley, "In the Field, Lap-Top Units Get Data, Print Proposals," *The Wall Street Journal*, September 13, 1985, p. 25.

3. Betsy Simnacher, "Bulletin Boards for Better Business," *Link-Up*, June 1984, pp. 32–34.

8

Home Automation Systems

OVERVIEW OF THE AUTOMATED HOME AND MAINTENANCE CENTER

A truly smart home is one that has an automated (computerized) operating system that controls its heating, cooling, lights, fire safety, and security, and can therefore provide residents with considerable savings. Savings in the area of lighting alone can be significant. Lighting, for example, accounts for about 40 percent of a home's energy consumption, yet most people neglect to turn off the lights when they leave a room. Automated systems take care of this problem by turning lights off when no movement is detected for 12 minutes, and turning them on when someone walks into a room.

Heating and air-conditioning sensors pick up environmental changes, such as increases in room temperatures when the sun shines on one part of a home. They enter the information into the computer, which then orders temperatures and airflow to be adjusted accordingly. This function is extremely important to the telecommuting environment, given the expected one-to-one person-to-terminal ratio expected in the near future.

Additional savings are provided by the smart home's automated security arrangement. When doors and windows are equipped with their own security devices, such as detectors which are connected to the computer and can be easily reprogrammed, there need be less concern for security.

Savings are also provided by a fire management system that detects heat or smoke and/or can be occupant activated, and which immediately signals a fire alarm. Should a fire break out, its location is reported to the central fire station where fire-fighting personnel are notified. At the same time, stairways and other critical passageways are lighted. In this way the fire is contained, while response time is minimized. The result can be a great savings in life and property.

To perform these functions, a home command and control center or system should monitor, diagnose, and, when possible, correct problems. Some control center functions include:

- A centralized computer system that efficiently handles home network control by monitoring the performance of a number of smart home systems. Monitoring can be performed as needed by remote control center operators, or automatically on a periodic basis. By also regularly monitoring system components, the center is informed of deterioration in their performance and can act before a malfunction causes a problem.
- Capability of testing components and diagnosing their defects or potential defects. Operators who test systems remotely may have video display terminals available through which they can access all of the systems in the home. From their terminals, the operators can investigate the status of system software and hardware, and directly perform a variety of diagnostic and repair functions. Whenever possible, network control centers provide operators with the means to solve problems involving defective components.

Whatever the transmission media (telephone or CATV) used by the network or the equipment involved, it is necessary to identify problems and then take steps toward remedying deterioration or failure in performance. In order for the home control center to operate effectively, it must determine the effect that the failure of any one component will have on the other home systems. In effect, the security system should "know" what the fire-system status is and act accordingly.

The "intelligence" of a home control center is directly related to the degree to which its functions are automated. In order to provide the highest quality service to users and to reduce their problem-related costs, it should periodically perform a series of tests on communications links and record the results. When a test result falls outside prescribed limits, a message appears on the printer in the control center and is also logged against the offending component in a large, computerized configuration database. This procedure enables operators to keep track of the history of each device or component.

It is desirable, however, to automate the process further. Even when test results are within acceptable limits, "symptoms" may appear that indicate to an investigator the probability of a defect surfacing within a determinable time. This function is in the realm of artificial intelligence and expert systems.

Even further automation is possible when advanced home control systems are installed at major nodes in the network. Advanced systems go into operation when the system finds, for example, that a device such as a heater or refrigerator shows potential for trouble. The device is then

Examples of Home Automation Equipment

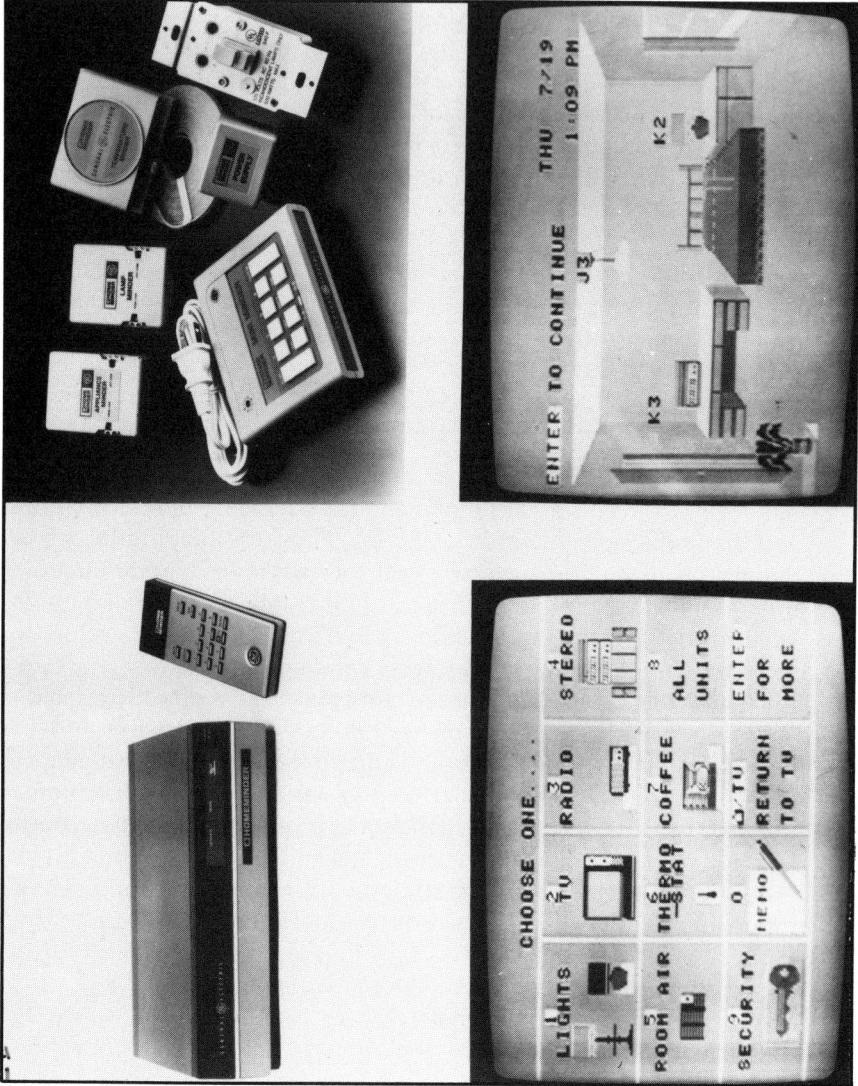

Courtesy Intelectron.

Home Automation Equipment

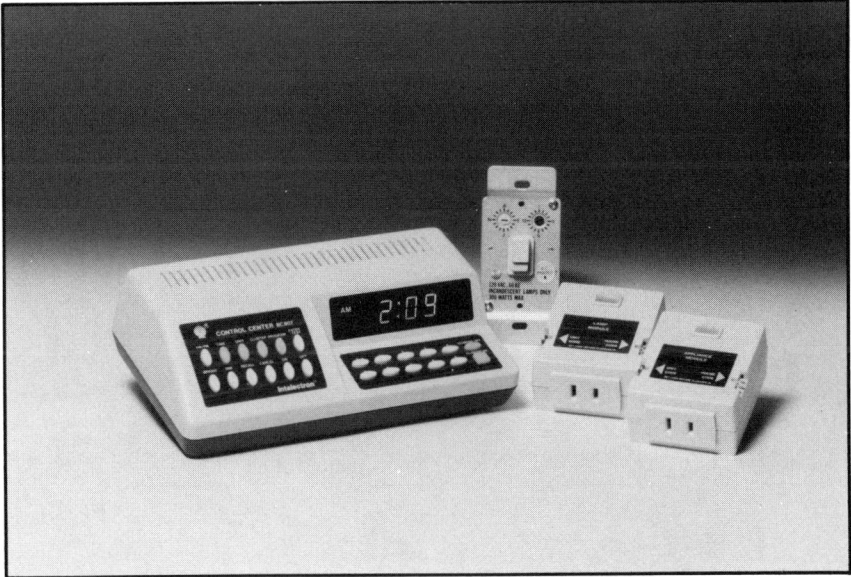

Courtesy Intelectron.

switched out of the home network, the remote repair center receives a description of the problem, and handles it. Using this process, no user would ever have a problem with the dishwasher or other appliance. The device would simply be taken out of service before it acted up, serviced without any interruption or degradation of network service, and returned to use in trouble-free condition.

Although it is difficult to give a strict dollars-and-cents accounting of control center cost benefits, an economic analysis of a typical problem will show on a qualitative basis why such a center is a worthwhile investment for any organization that maintains its own information network. To perform such an analysis, each home must determine the costs it incurs during each phase of a problem and the impact of one component upon another. The intangible benefits of shortening these events are improved home efficiency, lower maintenance costs, and reduced worry in regard to appliance or system upkeep.

The remote maintenance center provides a central site where thousands of homes, work centers, or offices can be monitored simultaneously and expertise can be concentrated for performing necessary repair functions. Thus, centralized network control can be justified from both an operational and an economic basis.

It is the personal computer (PC) with its established, generally accepted standards that will drive the development and integration of home appliances. With the proliferation of the personal computer and

---

Smart Home Computer System

Courtesy CyberLynx—Boulder, Colorado.

the increasing use of intelligent household appliances, we begin to address the reality of the smart home. Presently, many home devices (washers, refrigerators, heating-cooling systems, microwave ovens, etc.) are made with advanced microprocessors that enable them to perform self-diagnostics and programmable functions, and follow elaborate instructions, to name only a few activities.

Manufacturers are beginning to address the fact that these systems could interact and thus communicate with one another if they shared common standards. As a result, industry groups have formed task forces to address development of a common industry communications standard. It is unlikely that such a standard will emerge over the short term, however, since each company remains hopeful that the others will adopt its format. In addition, an industry standard might be required to provide for the inclusion of infrared, ultrasonic, and other technologies. The growing concern with standardization indicates that the day is fast approaching when the intelligent appliances of individual manufacturers will be produced to communicate with each other.

It would be possible for each appliance manufacturer to have its own proprietary standard for its own devices but at the same time structure those devices to interface with the dominant personal computer operating systems (OS) such as Apple DOS, MS–DOS®, and/or UNIX™. In the smart-home setting, this would allow the PC to monitor security devices, determine the optimum time for the water heater to be turned on, order food and other supplies after checking preset inventory levels, turn on the clothes drier when residents are asleep or out, send and receive mail, and handle other assignments that require home management time.

Ultimately, these services would:

- Save or manage residents' time more effectively.
- Reduce energy, food, or other costs.
- Allow for improved options in family entertainment, business communications, and education.
- Simplify running the home.

How will these systems emerge? Various smart-home manufacturers (see the Resource Guide) are rapidly developing a range of products. Much of their technology is based on the existing electrical power wiring system rather than the telephone wires in the home. Since most local building codes require that electrical outlets be placed on nearly every wall in every room, this usually offers ample connections for placing smart devices. Thus, the electrical system is used to transmit communications control signals to devices that are plugged into outlets throughout the house. Simple digital coding is used to tell devices what to do, such as turn a light on or off, detect a security break, or turn on a burglar alarm. This type of technology is generally called carrier-controlled, power carrier, or powerline carrier systems.

Newer technologies are emerging that incorporate additional control systems. Some of these use wireless radio frequency (RF) sensors and controls, much like remote garage-door openers. Other systems utilize infrared waves, which may have fewer problems, including interference from radio frequency systems. However, infrared is often limited to the confines of a room and rarely works around corners. Infrared systems can be incorporated into a wide range of devices and already exist in water faucets and toilets.

One of the most exciting telephone technologies that may play an important role in telecommuting is cellular radio. As mentioned in the electronic communications section of this book, cellular radio offers citywide and nationwide telephone service. While this application may not be evident for the home, data communications traffic can reach the home, and home appliances can have data receivers.

Subcarrier frequencies on normal FM band radios offer the same potential. These are currently being used to transmit stock market and weather information to people with decoders. Addressable cable television systems offer some of the same capability. For example, cable subscribers could tie all of their home appliances into the citywide cable system. They might then use the cable system computer to program or instruct the home appliances to perform certain functions. Then, at the appropriate time, the computer would send signals activating these devices. This approach in conjunction with the cellular telephone approach would allow office or work-center employees to turn on home devices, from the car or from another city.

ROBOTS

The next generation of smart homes is expected to include robots. These useful helpers have increased in numbers and functions over the past few years, particularly home robots that have mobility as a key feature. Some robots are able to avoid obstacles, speak, and fetch the newspaper. However, robot vision, still in its early developmental stages, is often crude, slow, unreliable, and very expensive. With the decrease in cost for video cameras and other new visual recognition technologies, vision systems for robots will no doubt emerge that are not only trainable but learn on their own.

This technology will reside in software systems that require artificial intelligence, expert systems, and learning processes that allow a robot or other device to "know" what to do rather than to merely act. For example, the simple act of opening a can of soup, making a telephone call, or vacuuming the carpet are all simple to a person. But it is very complicated to design a machine that can do any one much less all of these functions. Since each home has a different floor plan, stairs, and furniture layout, the robot needs to know where each obstacle is located in order to navigate through this maze. Stair climbing technology remains a problem that must be overcome before the robot can be free to roam throughout a house. While small toys are capable of moving across the floor easily and avoiding obstacles, few toys can climb or descend stairs.

Robot technology is being driven by the advent of microprocessors much like the personal computer. Some robots already have personal computers on board. Advanced robotics suggests that robots will have many (perhaps even hundreds of) microprocessors, each performing specialized functions in much the same way as human muscle groups.

PC "brains" may one day act to coordinate or network those group functions. This approach would reduce the software development required for robots because each activity would be under its own computer system and not interrupt other computer functions to work. The process of isolating problems and errors is also reduced because each activity can be identified with a small number of microprocessors. These tiny computers can be used for a wide variety of functions, including for force and touch sensors, recognition, security, safety, and information gathering (photographing or recording scenes for later use, such as keeping an eye on the baby).

One of the more intriguing concepts is connecting the robot with larger computers to "discuss" problems, learn new features, and be programmed to perform new tasks. It is envisioned that robots will have

Courtesy General Robotics Corporation, Golden, Colorado.

FIGURE 8-1 A Possible Design for a Household Robot

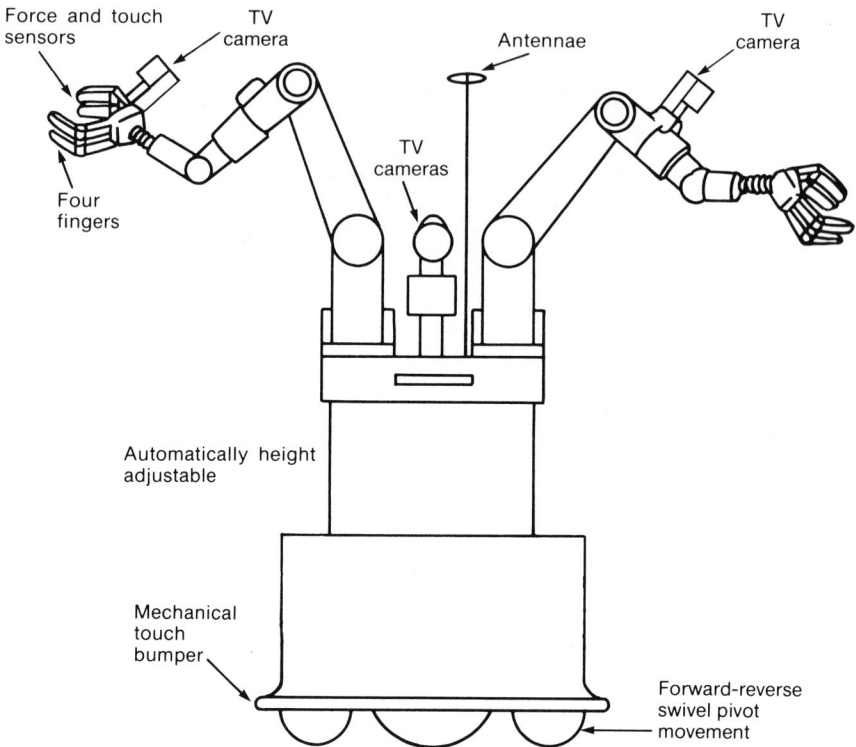

their own information bases as well as electrical outlets. Experimental robots are already capable of finding electrical outlets to recharge their batteries when they are low. See Figure 8–1.

Robots might be priced or marketed on the basis of their "IQ" (intelligence quotient) or job functions. For example, Robot 1 might be able to walk and have limited watchdog functions whereas Robot 100 might be able to conduct a conversation on lunar mining. By purchasing plug-in modules, buyers could increase their robot's intelligence as the need arose. There might even be robot-of-the-month clubs that will offer the owners new functions each month.

The robot is just another form of smart appliance which will interact with other devices so they can function effectively in the home. The robot "brain" will direct incompatible devices, devices that are too old, or devices that have no communications capability. Telecommuters will use them to supplement housework, walk the dog, empty the trash, monitor children, and plan the menu.

Conclusion

There is no "cookie cutter" approach to integrating telecommuting technologies. Whereas many remote workers use nothing more than their kitchen table and a telephone to perform their jobs, others require elaborate video/computer/telephone automation systems with robots. As the reader has probably noted, this book emphasizes communications. This is because people in business (whether upper-level managers or payroll clerks) spend much of their time communicating. For the telecommuter, isolation appears to be the most limiting factor. Thus, the type of communications system used for remote work may be the most important factor to consider in designing an effective telecommuting system.

In summary, the most suitable business conditions for telecommuting combine management commitment, an ergonomic environmental setting, and appropriate technology.

9

Future Technologies for Telecommuting

INTRODUCTION

This chapter provides an overview of some technologies that are expected to impact telecommuting, although there are few standard technologies that govern the remote-work setting. Current indications are that the personal computer will have the most dramatic impact on telecommuting, yet thousands of telecommuters work effectively and efficiently on other devices.

Bell operating companies (BOCs) are probably carrying out the most active research programs in telecommuting at present. They believe that working remotely will be one of the more important applications of such advanced information technology as the integrated services digital network (ISDN). Although ISDN is generally considered to be only a business technology, when it is used for telecommuting, ISDN becomes a very technologically competitive tool which the Bell operating companies are well positioned to provide.

ON-LINE DATABASES

On-line services are huge information utilities or databases that offer vast libraries of data to subscribers. While most are modest bulletin board systems (BBSs), three of the major on-line companies are: H&R Block's CompuServe, Reader's Digest's The Source, and the Dow Jones News Retrieval Service. The following provides an idea of the types of on-line services available from these companies by using a personal computer:

- CompuServe:
  a. Special interest groups (SIGs) comprise more than 50 on-line "clubs" that meet to exchange views and information on almost any subject.
  b. Encyclopedias:
     Grolier's Academic American, a 9 million word database that has more than 29,000 subject entries.
     World Book Encyclopedia, 31,000 subject entries and more than 10 million words.
  c. Business services providing: Figures on more than 9,000 securities that are updated every 20 minutes throughout each trading day, financial information on thousands of major publicly held companies, and current and historical information on more than 40,000 stocks, bonds, and options; specialized reports on commodities and today's economy, implications for the future, and financial commentaries from the nation's leading business and news publications.
  d. Information on demand (IOD): Provides access to professional research services. Will investigate, for a fee, any topic of interest in the news media and professional journals. Performs market and technical research and provides, at a special rate, English translations of technical material.
- The Source
  a. Electronic mail (service): Each subscriber has an electronic mailbox addressed by account number, that they can use nationally. The service can be used while one is traveling on the road to communicate with the home office and field representatives, to stay in touch with clients and suppliers, or to send messages to friends and relatives.
  b. Computer conferencing: A conference may include from 2 to 200 participants who engage in business meetings or committee discussions on any topic chosen. (See Chapter 7.)
  c. Chat (system): Fellow subscriber Source members may meet and communicate electronically, often for less than the cost of a telephone call.
  d. News and sports: Associated Press Videotex service provides 250 daily dispatches of national and international sports, business, and weather news.
  e. Retrieval and research: Subscribers may order electronically any book in print. Summaries are also available from 27 leading business publications such as *Forbes, Venture,* and *Harvard Business Review.* Customized research may be ordered for a fee.
- Dow Jones News Retrieval Service: Offers more information than any other on-line vendor in America. Subscribers can choose

from among 26 databases in three categories: business and economic news, financial and investment services, Dow Jones quotes, and general news and information. The company provides quotations with a minimum 15-minute delay during market hours. In addition, it provides a year's report on daily volumes, high, low, and closing figures, monthly summaries covering the past five years, and quarterly summaries for the past four years. On-line information features entire texts of *The Wall Street Journal* and *Barron's* for the previous several months. See Figure 9–1.

On-line services can be of great value to telecommuters, particularly those services that place information at the telecommuters' fingertips. To receive such on-line services, telecommuters require only a personal computer, a modem, a telephone line, and a subscription to the desired service. The service then provides a telephone access number, user code, and password.

Some of the specific business information applications for on-line services are:

Category of Information	Examples	Telecommuting Use
Training, education, information needed to respond to a specific function	Learning "how to"	High
General information	Newspapers	Fair
Corporate information	Bulletin board	High
Action items	Numbers, facts	Good
Information leading to a specific activity	Surveys	Good
Research-support information	Same as above	Good
Transaction/purchase replacement for telephone calls	Yellow pages, stock quotes, most items for sale	Good, limited by fair graphics

SOURCE: Cross Information Company.

VIDEOTEX

Videotex systems, called viewdata and teletext, are other interactive systems that allow users to send and receive text and graphics via either a personal computer or a keyboard and decoder unit that are connected to a television set. These systems generally present information on a television screen rather than on a computer monitor. The service is provided either "over the air" by television broadcasters or by wire from CATV (Community Antenna Television) cable operators.

FIGURE 9-1 News Retrieval Guide

SUBJECTS / **DATABASES** / **SERVICES**

SUBJECTS	//DJNEWS Dow Jones News	//UPDATE Weekly Econ. Update	//WSJ Wall Street Journal Highlights	//CQ Current Quotes	//CQE Enhanced Current Quotes	//DJA Historical DJ Averages	//FUTURES Dow Jones Futures Service	//HQ Historical Quotes	//RTQ Real-Time Quotes	//TRACK Dow Jones Tracking Service	//TEXT Wall Street Journal Full Text	//TEXT Dow Jones News Archive	//DSCLO Disclosure II[1]	//EARN Corp. Earnings Estimator	//FORBES Forbes Directory	//KYODO Japan Economic Daily	//MEDGEN Media General	//MLYNCH Merrill Lynch Research
Commodities	●						●											
Subsidiaries and Private Companies											●	●						●
Public Companies	●				●	●		●	●	●	●	●	●	●	●	●	●	●
Foreign Companies										●	●	●				●		
Canadian Companies	●									●	●	●						●
Industries	●									●	●	●		●	●	●	●	●
Government	●		●							●	●	●				●		
Securities Markets	●		●	●	●	●		●	●	●	●	●				●		●
Economic News	●	●	●	●	●				●	●	●	●			●	●		
Forecasts														●	●			●
General Information			●		●				●	●	●	●						
Subscriber News and Instructions																		
Airline Fares and Schedules																		
Mail Service																		

SUBJECTS	//MMS Economic Survey	//OAG Official Airline Guide	//SP Standard & Poor's Online	//DEFINE The Words of Wall Street	//ENCYC Encyclopedia	//MEDX Medical and Drug Reference	//MENU Master Menu	//MOVIES Movie Reviews	//NEWS World Report	//SCHOOL Peterson's College Selection Service	//SPORTS Sports Report	//STORE Comp-U-Store On Line	//SYMBOL Symbols Directory	//WTHR Weather Report	//WSW Wall Street Week	//MCI MCI Mail	//INTRO Intro
Commodities																	
Subsidiaries and Private Companies																	
Public Companies			●										●				
Foreign Companies																	
Canadian Companies													●				
Industries					●								●				
Government					●				●				●		●		
Securities Markets															●		
Economic News	●														●		
Forecasts	●														●		
General Information			●	●	●	●		●	●	●	●	●		●			
Subscriber News and Instructions							●						●				●
Airline Fares and Schedules		●															
Mail Service																●	

	DOW JONES BUSINESS & ECONOMIC NEWS	DOW JONES QUOTES	DOW JONES TEXT-SEARCH SERVICES	FINANCIAL & INVESTMENT SERVICES	GENERAL NEWS & INFORMATION SERVICES	MAIL SERVICE AND FREE CUSTOMER NEWSLETTER[3]
Prime Time	$1.20	$.90	$1.20	$1.20	$.60	
Non-Prime[2]	.20	.15	.60	.90	.20	
	Membership savings Executive—33⅓% anytime; Blue Chip—33⅓% non-prime time.					

NOTES [1]Disclosure II has an additional access fee of $5 for each company search in prime time, or $2 in non-prime time. [2]Non-prime time begins at 6:01 p.m. local time, and ends at 4 a.m. Eastern, 3 a.m. Central, 2 a.m. Mountain and 1 a.m. Pacific times. These prices apply to 300 baud transmission. News/Retrieval also is accessible at 1200 baud, in which case rates are 100% higher. [3]There are no online connect charges for MCI Mail. For message delivery fees, see //INTRO

The News Retrieval Guide is designed to help subscribers take full advantage of the complete News Retrieval service. For a total information picture of your subject, use this guide—which can be detached and kept close to your terminal—as a quick and easy directory to the databases you'll want to check.

Courtesy Dow Jones News Retrieval Service®.

The most attractive aspect of videotex systems is their ease of use. Simple menus and prompts enable the user to access information from a videotex database. The information, whether text, charts, or pictures, is displayed in a single-frame format, much like a page in a book. This procedure for accessing information requires no knowledge of computers or programming. Videotex systems are both easy and fast to operate and therefore use only a small amount of the supplier's computer time, slashing computing costs.

In the United States, videotex was first marketed experimentally to individual homes rather than to businesses. Expectations were that consumers would respond enthusiastically to the opportunity of banking and shopping via termials connected to their television sets. It has not worked out that way. Results have prompted industry watchers to suggest that it is within the corporation that videotex will see substantial use. In fact, many videotex vendors have focused their efforts on packaging this technology as a new information system designed for corporate use and refer to them as private, in-house videotex systems.

Videotex is used in a variety of ways. For example, telephone companies currently employ them for internal business use. One company has purchased two systems in order to acquire a working knowledge of the technology. It based the decision on data that indicates videotex will carry this type of network traffic in the future and on the realization that it was preferable to understand videotex through hands-on expertise—as opposed to reading a written description.

One system was used almost exclusively by senior management personnel. It currently features company and marketing news, a list of area seminars and conventions of interest to the company's major accounts, electronic messaging, and a list of news stories of interest to the company's department managers. Many vendors visualize businesses using their videotex systems to provide employees with an inexpensive medium for the rapid dissemination of timely information.

Videotex can also present an on-line information service used to gain access to databases throughout the world by accessing the company's or another host computer. In such an instance, the telecommuter or company pays for on-line services on a usage basis.

Videotex is sometimes called "teletext" in business where the system is accessed through personal computers and offered to telecommuters. The teletext system could provide users with a menu of all the company departments, providing easy access to information on each one. It could also be used as an information distribution medium from managers to telecommuters, among telecommuters, and from telecommuters to visitors. For example, a telecommuter could access information on company policy and procedures or about other telecommuters, eliminating the time-consuming bother of searching and collecting that information from various organizational departments.

The novel feature of most teletext offerings is that they converge on the concept of the personal computer (PC) as the workstation. However, certain systems are based on the use of a dedicated terminal. For example, one system consists of a wireless keypad and a controller unit that connect to a standard television set and telephone, transforming the terminal into a videotex database access system. Another system offers both a television-connected terminal and a stand-alone terminal monitor. The advantage of PC-based videotex systems is that they allow potential users to take advantage of their installed terminals, eliminating the cost of purchasing additional hardware.

The expected videotex revolution in the corporate environment will affect the development of a wide range of businesses, especially but not exclusively in the information and communications industries. Videotex can create an information retrieval system for mass market use which avoids the limitations of existing computer systems. It is generally known for providing home or business connections to large consumer-oriented computer databases via "wire" services such as cable television, the telephone, and electric power. As such, videotex may eventually become the electronic equivalent of the Yellow Pages. See Table 9–1.

In addition, anything from electronic mail to electronic funds transfer could be accessed through the viewdata system, another type of videotex. Thus, the players in this rapidly emerging market are telephone companies, newspapers, broadcasters, electric utilities, cable television operators, manufacturers, and suppliers.

TABLE 9–1 Potential Future Applications of Videtex Systems

The following is a sample of potential services that could be provided by on-line databases and video/teletext systems:

Advanced applications and programs	Treasury bill auction reports
Business and finance	UPI/AP/Reuters general wire service
Daily news and features	Washington hotline
National real estate locator services	International and domestic news
New York Times business data	Business and financial news
Travel agent and reservation services	Sports news
Money-saving buying service	Local government information data bank
American Stock Exchange bond prices	Business research services
Federal Land bond prices	Voter information and license information
New York Stock Exchange	Real estate locator services
Treasury bill values	Mailgram receiving services
World Bank bond prices	Restaurant listing and reviews
Commodity prices and futures	Automobile repair information
Earnings reports	Income tax information
Financial commentary	Social security information
Foreign exchange rates	Automobile renting/leasing
Financial news and headlines	Lodging review and reservation
Stock market averages	Bill-paying services
Stock quotations—news	

SOURCE: Cross Information Company.

Conclusion

Educators and businesses have for many years been communicating via data communications packet networks from all points of the globe to massive research database systems. Videotex technology extends this concept for use by the average household or small business by reducing the cost to affordable, available, usable levels. The office of the future is a logical place for videotex services, with the viewdata system providing a communications/information link between the home and office as well as a link between branch offices of a company.

INTEGRATED SERVICES DIGITAL NETWORK (ISDN)

ISDN is characterized as an end-to-end digital network that provides digital services that range from very low-speed telemetry to high-speed digital facsimile to very high-speed (1.5 mbps) video teleconferencing. In the not-too-distant future, a fully digital voice and data switch at the heart of the intelligent home will also serve the telecommuter.

The home or business subscriber will access the ISDN(s) via a standard modular telephone jack (interface) where voice and/or data traffic will be multiplexed into a digital bit stream and sent to its prescribed destination. A neighborhood or building multiplexer-concentrator will take advantage of the fact that individual lines are used only a small fraction of the time, even during the busiest hours, and allow the access lines to use any available time slots on the digital carrier system.

A key factor in the development of ISDN is the availability of twisted wire pair(s) in homes and businesses. The switching function can be provided in a number of ways, by a private branch exchange (PBX), at a centralized telephone company office (CO), or by combinations of PBXs and COs. Although ISDN is in its early developmental stages, it provides a wide range of network services over a common set of digital network facilities. These include voice, data, images, music, and video transmission services.

In the United States, the public-switched telephone network has been evolving toward an ISDN, and international standards committees (such as CCITT) are working to develop a worldwide system. The original impetus was to use electronics as a way of saving on the use of copper wire and duct space in the metropolitan area interoffice trunk environment. Digital multiplexing was introduced into the telephone network to reduce the cost of interoffice trunks in metropolitan area networks.

The ISDN application known as the North American digital T1 network system allows 24 conversations to be multiplexed (the European system provides for 30 voice channels). In the T1 application, analog-to-digital conversion occurs in equipment (channel banks) that is located in each switching office, and repeaters are employed at regular intervals along the cable route. By installing T1 systems, telephone companies are able to avoid the cost of installing expensive copper pairs for each individual interoffice trunk.

This economic advantage is accentuated in intelligent homes and buildings because transmission is maximized while the cabling required is minimized. Installing new cable can be very expensive in homes and buildings, especially when new ducts are required.

The development of the T1 system utilizing cable pairs was followed by that of comparable systems which use other transmission media such as coaxial cable and microwave radio. High-speed data transmission became a possibility because certain media allow much higher transmission speeds. This in turn led to the emergence of data communications and the demand for combined voice/data transmission.

The basic building blocks of an ISDN network are digital multiplexers and digital switches. The technical advantages of such networks are:

- Ease of multiplexing for the sharing of facilities.
- Ease of signaling.
- Use of digital technology for communications.
- Integration of transmission and switching.
- Ability to operate at low signal-to-noise (interference) ratios.
- Signal regeneration for long-distance transmission.
- Accommodation to other services.
- Performance monitoring for maintenance.
- Ease of encryption for privacy and security.

See Figure 9–2.

The most important advantage of the evolving ISDN system is its accommodation to other services such as nonvoice message services. This allows transmission of data, video, and practically all other forms of communication. Depending on a number of factors including the adoption of standards, competing systems, and the regulatory environment, several ISDN systems and standards are likely to evolve in the United States.

There remains a question of future demand for many of the services an ISDN would support. Videotex (mentioned earlier) is one such service forecast for ISDN. It should be noted that a major advantage of the ISDN concept is that it is not service specific. It also can be used for telemetry, sending interactive data, images, voice, and video traffic, although the switching technique used (e.g., circuit, message, or packet) would vary depending upon the statistical characteristics of the traffic. The host of potential applications ranges from games to services for government, banking, training, and the media. The bulk of ISDN usage by business may be in the transmission of large amounts of data traffic from computer to computer, at high speed and on demand.

FIGURE 9-2 ISDN Architecture, Functional Diagram

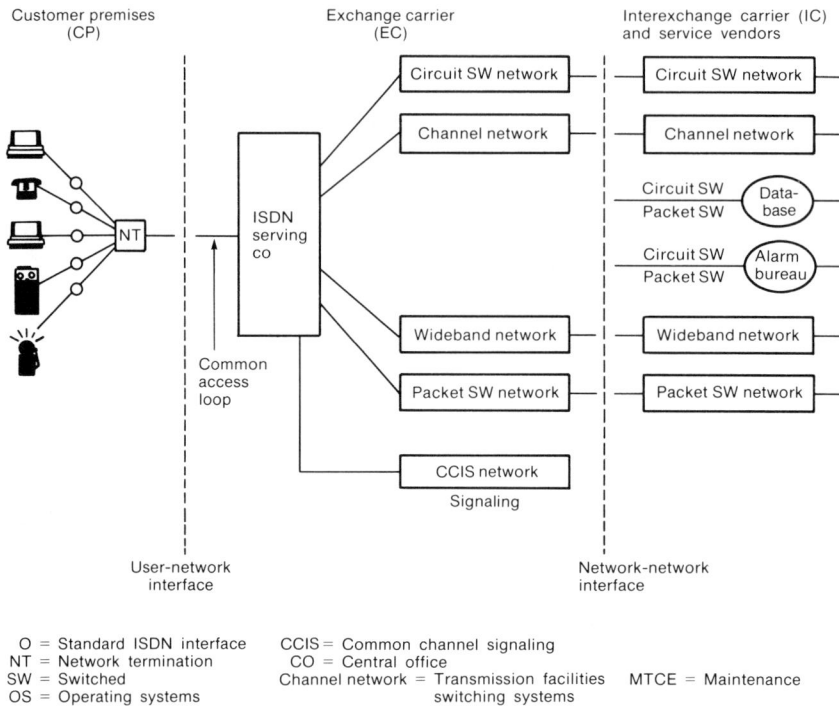

O = Standard ISDN interface   CCIS = Common channel signaling
NT = Network termination   CO = Central office
SW = Switched   Channel network = Transmission facilities   MTCE = Maintenance
OS = Operating systems   switching systems

In summary, ISDN has important implications for intelligent homes and buildings. Intelligent buildings using an ISDN can communicate among themselves and to home telecommuters. ISDN will also play a critical role in the development of teleports. A PBX or central office provides local processing and switching for communications. It then channels them through the connected ISDN local area network to a teleport or other network gateways and on to other ISDN domestic and international networks. ISDN is believed to be the next step in the evolution of digital technology.

Certain technologies, such as artifical intelligence and expert systems, may play a limited role in the home and for the telecommuter. Smart appliances, robots, and PC-operated "expert" systems that can manage the home in a cost-effective and pleasant manner would be highly desirable for people who want more time for career work as well as for recreational activities. Expert systems could also be used to train children, retrain and update adults on new managerial trends and current news patterns, or in handling household affairs.

Artificial intelligence (AI) systems could also be used to diagnose health problems, suggest preventative exercise, diet, and activities, or to play mind games as well as to train robots, manage the home command center, pay bills, and—in the future—anticipate problems that a robot could handle.

ARTIFICIAL INTELLIGENCE AND TELECOMMUTING

Author's note: The relationship of artificial intelligence to telecommuting is a complex subject. Thus, the following discussion is relatively abstract, although the authors have tried to make it "user-friendly."

Fundamental to artificial intelligence research is the fact that there are *many* ways to find information on a subject. Our basic precept is that one's way of thinking affects the way one approaches, analyzes, and solves a problem. As a result, people need power tools that can be adapted to individual ways of thinking. And as we advance from programming to knowledge processing, there is not just one approach or one system that offers a single avenue to problem solving. As can be seen in a brainstorming session, there are many ways to arrive at the same conclusion.

Ergonomics, the study of mind/machine interface discussed earlier, is typically approached from the perspective of "making the machine adapt" to the user, or "keeping a software program simple, so the user is not frustrated." In telecommuting, where much of the communication and interaction takes place via a PC or other teledevice, user ergonomics will be a key element. There is another approach to erognomics that has for the most part been overlooked, however. It is expressed by the question, "How can the user be made more aware of the power of the machine?"

The situation is well illustrated in the following example. Say that you, the user, have a problem and want to consider different aspects of it. Your computer is friendly enough, but you are stymied because you don't know which subject areas are related and should be researched. As a result, you cannot even browse through the information in the friendly machine to solve your problem.

In the future, telecommuters will be able to access remote systems that can probe the larger context of the problem. These range from expert, decision support, and knowledge networking systems to artificial intelligence systems. How will they help the user? "Problem solving in artificial intelligence requires that a representation of the problem be available so the program concerned can (be) usefully exploited—whether or not it is thought of at the level of the machine code," according to Margaret A. Boden in her seminal work, *Artificial Intelligence and Natural Man.*[1]

Boden further emphasizes, "We are particularly interested in how the structure of the plan can aid thinking by suggesting methods of inference that can lead economically to a solution. In other words, it is important to be able to see "the big picture" in order to know how to get the information that is needed to solve a problem. Boden is highlighting the necessity of envisioning the scope of a problem in the new systems before approaching its solution.[2]

However, much of the knowledge we possess and many of the ways we solve problems are not linear. They are "illogical," subjective, intuitive, and experiential, claims research specialist Yoneji Masuda.[3] They include:

- Nonlinear complex problems such as discrimination.
- Unanticipated problems, e.g., "Love Canal" situation.
- Heretofore unknown problems, e.g., AIDS.

Traditional learning tools restrict us in the way we approach this type of problem solving. In telecommuting situations, complex or unknown problems can be networked among telecommuters. In designing artificial intelligence systems that would be useful to business, one of the most difficult problems is understanding how humans think. Here researchers ask how we organize our ideas, or link one area of knowledge to another that is apparently unrelated.

An indication of the way in which our minds work is illustrated by the way we read a newspaper. We skip around, read something intently for a moment, then jump to another article once we are bored. How do we decide what to read? Asked in another way, why do we choose some articles and skip others? According to any one person's perspective, there is an amazing amount of trivia contained in the newspaper, a lot of "useless information content."

The concept of the newspaper goes to the heart of the way in which facts are gathered, presented, and read. Do you, for example, sit down and say, "I want to know about the Middle East, the space shuttle, or Wimbledon"? Clearly not. People skim through the newspaper, focus on articles they find interesting, then move on to other stories. As a result, our "information-eating habits" could include "gobbling up" some lead stories, cartoons, and editorials, while ignoring others.

With great insight into the man-machine interface relationship, John Seely Brown, a specialist in artificial intelligence research and computer-based tutoring systems, offers the following insights: "As equipment gets more complex . . . the user will feel enslaved by his technology. . . . Our mandate is to open up new areas of man-machine communications."

In forecasting our ability to manage complex technology, Brown opines, "By 1990, what's available in the computer marketplace will be constrained not by technology but by what people are capable of understanding."[4]

TELECOMMUTER IDEA EXCHANGES

An idea can be linked by electronic hardware to other ideas to form an idea exchange network within which many people can work. Using an idea exchange, people have applied particular types of technology in various ways. In very simple graphics, idea networking charts the user's position at a given moment. If the user wants to move to another point, the system will chart that point. This process is called linking and would be diagrammed on the computer screen to look like this:

An Idea Exchange Concept

```
                              CHART
                              ‾‾‾‾‾
       EXCHANGE               EXCHANGE               EXCHANGE
       IDEAS                  IDEAS — — LINKS — — — IDEAS
       IDEAS                  IDEAS                  IDEAS
       IDEAS — — LINKS — — — IDEAS                  IDEAS
       IDEAS                  IDEAS — — LINKS — — — IDEAS
       IDEAS                  IDEAS                  IDEAS
```

SOURCE: Courtesy Cross Information Company.

Telecommuting is a networking or linking process. It must be flexible as well as capable of "capturing" different telecommuter positions in separate locations and consider a variety of approaches to thinking. The electronic network could be used by a group of telecommuters who are writing a product documentation manual together. As the group develops its presentation, each product feature is charted and linked, and also organized and edited by the various telecommuters without ever requiring them to meet face-to-face.

TELECOMMUTING AS A DECISION SUPPORT SYSTEM

Remote workers link their work and ideas to the office just as managers link the many related people, issues, data, and facts that pertain to their job objectives. Hence linkage can help management make hard decisions such as those pertaining to short- or long-term investment, support for research and development versus present product positioning, or personnel raises versus plant improvements.

Just as the telecommuter may require information from computer databases in dealing appropriately with a new situation, the manager might tap the following database areas and link them as a network in deciding on a new location for a plant:

- Data on land costs.
- Labor costs.
- Existing transportation facilities.

- Transportation charges.
- The nature and disposition of local governments.

EXPERT SYSTEMS

Expert services, available on-line to a telecommuter network, provide access to a wide variety of databases. In simple terms, an expert system (ES) is a computer software system that organizes knowledge and rules or procedures to solve problems. If properly designed and maintained, an expert system can perform at or near the level of a human expert. The key issue is that an expert system is a machine that can only reflect the background and limitations of its creator/designer and the skill and knowledge of the people who use it.

The development of database retrieval services, such as on-line systems and viewdata/videotex, are growing rapidly. People-oriented systems such as electronic mail, computer teleconferencing, and on-line expert services are also emerging as powerful telecommuting tools. In deciding which strategies and opportunties to pursue, telecommuters will have the option of utilizing both personnel and database resources.

In conclusion, the application of advanced information technologies such as networking, idea exchanges, expert systems, artificial intelligence, and others on the horizon, will change the way business is conducted in the office or from the home. As telecommuters network with other employees, their support systems (such as expert systems) will provide them with "executive helpers" and/or "power tools" that allow them to be more productive in their jobs.

TRENDS IN TELECOMMUTING

Although many people consider telecommuting to be trendy, the technology for working remotely has existed for many years. Telecommuting is, in fact, a social rather than a technological phenomenon.

The following trend projections are included to give the reader a glimpse of technology that is expected to develop over the next five years. Note the number of technologies that are in place today in Table 9–2. With this benchmark as a guide, readers may enjoy noting how closely the following technology comes on schedule. It may also be useful for readers to keep these technologies in mind when planning for the future or waiting for even newer technology to emerge.

Information Technology—1986:
- First-generation intelligent buildings/teleports.
- Rapid growth in lap computers.
- Modular "windows" teleconferencing systems.
- Smart teleterminals emerge.
- First-generation cellular radio.

TABLE 9-2 Estimated Household Penetration by Product (as of June 1985)

All television	98%
Color TV	91
Monochrome TV	70
Projection TV	2
VCR	23
Programmable video games	25
Home computers	13
Audio systems	87
Compact	51
Component	38
Telephone answering devices	7
Cordless telephones	11
Home radios	98

SOURCE: Electronics Industry Association.

- PC computer-aided-design (CAD) office design systems.
- Gigabit per second fibre optics.
- Early stages of telecommuting.

Information Technology—1987:
- Portable PC-based video teleconferencing.
- "Expert systems" building technology.
- Standards adopted for local area networks.
- "Designer" environments—Gucci™ office furniture.
- Office robots and smart "chip" chairs.
- Sketchboard personal computers.
- Biocomputing technologies.

Information Technology—1988:
- Light-based computing emerges.
- Gigabit personal computers.
- Intelligent bypass technologies.
- Smart thought-processing software.

Information Technology—1989:
- Integrated PC-cellular office systems.
- Radio-controlled biological implant technology.
- Personal image support Bioenvironments.
- Corporate strategist gaming systems.
- Wearable offices—smart "chip" clothing.

Information Technology—1990:
- ISDN intelligent buildings and home environments.
- Dick Tracy cellular systems.
- Extremely large-scale optical storage—10^{15} bits.

Information Technology—1991:

- Worldwide networked intelligent buildings.
- Laser-panavision war rooms.
- "Watchman" management systems.
- User-programmable biological implants.
- 100 gigabit per second fibre optics.
- AI—HAL9000 smart buildings.
- Twenty million people telecommuting worldwide.

The preceeding technologies and issues are expected to impact the remote workplace, American lifestyles, and the way business is conducted. As we noted earlier, many people believe that there are no limits to technological progress, only the user's limitations in applying it. Table 9–3 gives us a glimpse of future projections for various types of products which may be used in telecommuting situations. Of course, one must remember that technology changes faster than we can anticipate, while society changes much more slowly than we can anticipate.

The next few years will see a dramatic increase in the numbers of companies and people who integrate telecommuting programs. However, remote work is not for everyone and people must be allowed to function in the type of environment they choose, whether the setting is a home, office, car, boat, train, or plane.

Conclusion

It is difficult to determine the direction in which telecommuting will evolve because there are so many technologies that can influence the remote-work picture. Fortunately, telecommuting will have meaning for those organizations that thoroughly understand their operating strategy and have experienced consultants to advise them on the best approaches to integrating the current technology into the remote-work setting—remembering all the while that telecommuting is a social process which is being advanced by technology.

NOTES

1. Margaret A. Boden, *Artificial Intelligence and Natural Man* (New York: Basic Books, Inc., 1977), p. 333.

2. Ibid., p. 336.

3. Yoneji Masuda, *The Information Society* (Tokyo: Institute for the Information Society).

4. Daniel Goleman, "The Human-Computer Connection," *Psychology Today*, March 1984, pp. 20–21.

TABLE 9-3 Total Factory Sales of Consumer Electronics by Product (in millions of dollars)

Product	1983	1984	1985 (projected)	1986 (estimated)
Video:				
Color TV	5,002	5,538	5,345	5,120
Monochrome TV	465	419	312	278
Projection TV	268	385	410	488
Total VCRs	2,162	3,585	4,846	4,918
Color video cameras	303	355	307	261
Video disk players	81	45	*	*
Videocassette players	n.a.	n.a.	23	20
Total video products	8,281	10,327	11,243	11,085
Audio:				
Audio systems	630	976	960	973
Separate audio components	1,268	913	1,100	1,200
Portable audio tape equipment	1,102	1,191	1,268	1,265
Home radios	565	661	560	560
Car audio	1,900	2,484	2,800	3,000
Total audio products	5,465	6,225	6,688	6,998
Other:				
Blank audiocassettes†	250	256	285	305
Blank videocassettes†	540	770	900	1,070
Videodiscs	150	90	*	*
Home computers	1,950	2,250	2,250	2,500
Home computer software	600	640	675	910
Programmable video games	760	165	65	35
Video game cartridges	1,400	765	200	100
Telephones	925	1,250	1,031	1,035
Telephone answering devices	190	230	277	334
Total other	6,765	6,416	5,683	6,289
Grand total	20,511	22,968	23,614	24,372

*EIA policy precludes publishing estimates where one company represents 50 percent or more of the market.

†Includes sales to consumer distribution only.

SOURCE: Electronics Industry Association.

10

Telecommuting Resource Guide

This Resource Guide is designed to provide workers either at home or in the office with useful information for leading up to as well as enhancing the telecommuting process. The Resource Guide contains information on the following subjects:

- Telecommuting conferences.
- Telecommuting support organizations.
- Telecommuting newsletters.
- Telecommuting books and related issues.
- Smart home products.
- On-line database services.
- Matrix of audio bridging service options.
- Telecommuting business case.
- The grapefruit diet.

TELECOMMUTING CONFERENCES

Telecommuting Technology Conference

Sponsored by:
Cross Information Company
1881 Ninth Street, Suite 311
Boulder, CO 80302-5151
Telephone (303) 444-7799

Workshops and seminars are also available.

TELECOMMUTING SUPPORT ORGANIZATIONS

The Association of Electronic Cottagers (AEC)
677 Canyon Crest Drive
Sierra Madre, CA 91024
Telephone (818) 355-0800

The AEC is an association of full- or part-time telecommuters that has developed "The Electronic Cottage Bill of Rights." The first provision states, "Legislatures shall make no laws prohibiting freedom of opportunity to work in one's home with a computing and/or robotic device when that work does not interfere with neighbors' enjoyment of their own homes and communities."

The association provides members with:

- Personal contact with a national network of telecommuters who have similar interests.
- Full and affiliate membership that lends professional identity.
- Promotional and marketing support.
- Business services.
- Legislative monitoring and support for pro-telecommuting legal issues.
- "Helps aspiring cottagers find work at home or set up a computer-based business."

National Association for the Cottage Industry (NACI)
P.O. Box 14850
Chicago, IL 60614

This association serves the home worker in the following ways:

- As a clearinghouse for home and business information.
- By representing the interests of the home worker.
- By attempting to protect the rights of all people to choose their workplace.

The National Alliance of Home-Based Businesswomen
P.O. Box 95
Norwood, NJ 07648

This organization is concerned with the legal issues of working from home. It has written a model zoning ordinance to protect both home businesses and neighborhoods.

Center for Futures Research
University of Southern California
Graduate School of Business Administration
Los Angeles, CA 90089-1421
Telephone (213) 743-5229

Electronic Services Unlimited
142 West 24th Street, 10th floor
New York, NY 10011
Telephone (212) 206-8272

Cross Information Company
1881 Ninth Street, Suite 311
Canyon Center
Boulder, CO 80302-5151
Telephone (303) 444-7799

Provides consulting and research.

The National Alliance of Home-Based Businesswomen
P.O. Box 306
Midland Park, NJ 07432

Lift
Northbrook, IL

Lift provides a six-month training program for the disabled that is administered gratis via video and cassette tapes. Trainees are then hired as contract workers through Lift.

The Workstead Network
P.O. Box 29464
San Francisco, CA 94129

The Work-at-Home Special Interest Group (SIG)
CompuServe Network.

TELECOMMUTING NEWSLETTERS

Cottage Computing
12221 Beaver Pike
Jackson, OH 45640

The MicroMoonlighter Newsletter
4121 Buckthorn Court
Lewisville, TX 75028

Periodicals that carry telecommuting articles:

ABA Banking Journal	*Fortune*
Across the Board	*The Futurist*
Business Week	*Infosystems*
Computer Decisions	*Management Review*
Data Management	*Management World*
Datamation	*Modern Office Procedures*
Forbes	*Nation's Business*

Newsweek
The Office
Office Administration and Automation
Personnel
Personnel Administration

Technology Review
Time
U.S. News and World Report
Working Woman

TELECOMMUTING BOOKS AND RELATED ISSUES

Michelle Gouin and Thomas B. Cross, *Intelligent Buildings* (Homewood, Ill.: Dow Jones-Irwin, 1986).

James Weidlein and Thomas B. Cross, *Networking Personal Computers in Organizations* (Homewood, Ill.: Dow Jones-Irwin, 1986).

Thomas B. Cross and Marjorie Raizman, *Networking: An Electronic Mail Handbook* (Glenview, Ill.: Scott, Foresman, 1986).

Kathleen Kelleher and Thomas B. Cross, *Teleconferencing: Linking People Together Electronically* (Englewood Cliffs, N.J.: Prentice-Hall, 1985).

Etienne Grandjean, *Ergonomics of the Home* (London: Taylor & Francis, Ltd., 1973).

Alvin Toffler, *The Third Wave* (New York: Bantam Books, 1980).

Charles Garfield, *Peak Performance* (New York: Wm. Morrow, 1985).

Franklin Becker, *The Successful Office: How to Create a Workspace That's Right for You* (Reading, Mass.: Addison-Wesley, 1982).

Roy Mason et al., *Xanadu; The Computerized Home of Tomorrow and How It Can Be Yours Today* (Washington, D.C.: Acropolis Books, Ltd., 1983).

Jack M. Nilles et al., *The Telecommunications-Transportation Trade-Off: Options for Tomorrow* (New York: John Wiley & Sons, 1976).

Jeremy Joan Hewes, *Worksteads: Living and Working in the Same Place* (New York: Doubleday/Dolphin, 1981).

Paul and Sarah Edwards, *Working from Home* (J. P. Tarcher; distributed by Houghton Mifflin, Boston, Mass., 1984).

Marion Behr and Wendy Lazar, *Women Working Home: The Home-Based Business Guide and Directory* (Norwood, N.J.: WWH Press, 1983).

SMART HOME PRODUCTS

BSR, Ltd.
Route 303
Blauvelt, NY 10913
(617) 964-3210

Cybot, Inc.
12510 128th Avenue, NE
Kirkland, WA 98034

Energy Methods, Inc.
Butler Manufacturing Company
177 Main Street
West Orange, NJ 07052

Haikato Robotics Ltd.
1580 Lincoln Street
Suite 950
Denver, CO 80203
(303) 837-8477

Heath Company
Hilltop Road
St. Louis, MO 49085

Homebrain
Hypertek, Inc.
P.O. Box 137, Route 22 East
Whitehouse, NJ
(201) 534-9700

Homenet
General Electric Company
Television Business Division
Portsmouth, VA 23705
(804) 483-5000

Honeywell Communications and Systems Group
Honeywell, Inc.
1700 West Highway 36
Roseville, MN 55113

Hubotics, Inc.
6352-D Corte Del Abeto
Carlsbad, CA 92008

Intelectron (Brightmond, Ltd.)
1275 A Street
Hayward, CA 94541
(415) 581-4490

Microbot, Inc.
453 Ravendale Drive
Mountain View, CA 94043

Plug 'n Power
Radio Shack
1 Tandy Center
Fort Worth, TX 76102
(817) 390-3000

Smarthome I
Cyberlynx Computer Products, Inc.
4848 Sterling Drive
Boulder, CO 80301
(303) 444-7733

RB5X
General Robotics Corporation
18301 West 10th Avenue
Golden, CO 80401

Rhino Robots, Inc.
P.O. Box 4010
Champaign, IL 61820

ON-LINE DATABASE SERVICES

ACP Network Services
Division of Automatic Data Processing, Inc.
175 Jackson Plaza
Ann Arbor, MI 48106

Services: 35 textual and numeric databases covering finance, banking, economics, and investment.

Bibliographic Retrieval Service (BRS)
1200 Route 7
Latham, NY 12110

Services: Bibliographic and full-text files; including medicine, education, engineering, finance, and reference sources.

Chase Econometrics/Interactive Data
Subsidiary of Chase Manhattan Bank
486 Totten Pond Road
Waltham, MA 02154

Services: Historic and forecast economics and financial databases.

CompuServe, Inc.
500 Arlington Center Boulevard
Columbus, OH 43220

Services: Business Information Services (BIS), Consumer Information Services (CIS), and Executive Information Services (EIS).

Data Resources, Inc.
A division of McGraw-Hill
24 Harwell Avenue
Lexington, MA 02173

Services: More than 75 business, finance, and economic databases; Lotus 1-2-3 compatible.

Dialog Information Services, Inc.
3460 Hillview Avenue
Palo Alto, CA 94340

Services: More than 100 million bibliographic records. Abstracted articles from 10,000 journals: business, industry, chemistry, medicine, law, government, science, technology, energy, education, news.

Dow Jones Information Service
P.O. Box 300
Princeton, NJ 08540

Services: Business, economics, financial and investment, stock quotes (current and historical), and current news.

Mead Data Central
The Mead Corporation
9333 Springboro Pike
P.O. Box 933
Dayton, OH 45401

Services: 17 million full-text documents from 125 periodicals, newsletters, newspapers, technical and scientific journals, and newswire access.

NewsNet, Inc.
945 Haverford Road
Bryn Mawr, PA 19100

Services: 164 full-text newsletters; telecommunications, publishing and broadcasting, electronics and computers, taxation, energy, investment, and finance.

Quotron Systems, Inc.
5454 Beethoven Street
Los Angeles, CA 90066

Services: Information on domestic and international stocks, bonds, mutual funds, commodities, and options.

TRW Information Services
Division of TRW Inc.
505 City Parkway West
Orange, CA 92668
Services: Business credit information.

BioSciences Information Services (BIOSIS)
2100 Arch Street
Philadelphia, PA 19103
(215) 587-4800, (800) 523-4806

Fred Browne Associates
P.O. Box 3490
Santa Barbara, CA 93130
(805) 687-1140

BRS Information Technologies
1200 Route 7
Latham, NY 12110
(518) 783-1161, (800) 833-4707

Commerce Clearing House, Inc. (CCH)
Marketing Department
4025 West Peterson Avenue
Chicago, IL 60646
(800) 621-6224

CompuServe, Inc.
5000 Arlington Centre Boulevard
Columbus, OH 43220
(614) 457-8600, (800) 848-8990

Cuadra Associates, Inc.
2001 Wilshire Boulevard, Suite 305
Santa Monica, CA 90403
(213) 829-9972

DATA–STAR
D-S Marketing, Ltd.
Plaza Suite
114, Jermyn Street
London SW1Y 6HG England
44(1) 930-5503

DIALOG Information Services, Inc.
3460 Hillview Avenue
Palo Alto, CA 94304
(415) 858-3785, (800) 227-1927

Diogenes
12315 South Wilkins Avenue
Rockville, MD 20852
(301) 881-2100

Dun and Bradstreet Inc.
99 Church Street
New York, NY 10007
(212) 285-7669

E + E Datacomm
2115 Ringwood Avenue
San Jose, CA 95131
(408) 288-8880

EIC/Intelligence, Inc.
48 West 38th Street
New York, NY 10018
(800) 223-6275

Grolier Electronic Publishing
95 Madison Avenue
New York, NY 10016
(212) 696-9750

GTE Telenet Medical Information Network
12490 Sunrise Valley Drive
Reston, VA 22096
(703) 442-2500, (800) 368-4215

Hayes Microcomputer Products, Inc.
5923 Peachtree Industrial Boulevard
Norcross, GA 30093
(404) 449-8791

Healthlawyer
Office of Legal Communications
American Hospital Association
840 North Lake Shore Drive
Chicago, IL 60611
(312) 280-6594, (800) 621-6712

Information Access Company
11 Davis Drive
Belmont, CA 94002
(415) 591-2333, (800) 227-8431

Institute for Scientific Information
3501 Market Street
Philadelphia, PA 19104
(215) 386-0100, (800) 523-1850

Inter Company Comparisons, Ltd.
ICC House
81, City Road
London EC1Y 1BD England
011-441/250-3992

Interlink Press Service
777 United Nations Plaza
New York, NY 10017
(212) 599-0867

Jordan and Sons, Ltd.
Jordan House
Brunswick Place
London N1 6EE
011-441/-253-3030

Market Data Retrieval
Ketchum Place
Westport, CT 06880
(203) 226-8941, (800) 243-5538

Marquis Who's Who, Inc.
Data Products Department
200 East Ohio Street
Chicago, IL 60611
(312) 787-2008, (800) 621-9669

Mead Data Central
P.O. Box 933
Dayton, OH 45401
(513) 859-1611, (800) 227-4908

Menlo Corporation
4633 Old Ironsides
Suite 400
Santa Clara, CA 95054
(408) 986-0200

National Technical Information Service (NTIS)
Database Services Division
5285 Port Royal Road
Springfield, VA 22161
(703) 487-4807

NEDRES NOAA/NEDRES
E/AIX 3
3300 Whitehaven Street NW
Washington, DC 20235
(202) 634-7722

NewsBank, Inc.
58 Pine Street
New Canaan, CT 06840
(203) 966-1100

NewsNet
945 Haverford Road
Bryn Mawr, PA 19010
(215) 527-8030, (800) 345-1301

OCLC Online Computer Library Center
6565 Frantz Road
Dublin, OH 43017
(614) 764-6000

Paladin Software Corporation
2895 Zanker Road
San Jose, CA 95134
(408) 946-9000

PDS Sports
P.O. Box E
Torrance, CA 90507
(213) 516-6688

Pierian Press
P.O. Box 1808
Ann Arbor, MI 48106

Paperchase
Beth Israel Hospital
330 Brookline Avenue
Boston, MA 02215
(617) 735-2253

Public Affairs Information Service (PAIS)
11 West 40th Street
New York, NY 10018
(212) 736-6629

Salem Press/Magill Books
580 Sylvan Avenue
Englewood Cliffs, NJ 07632

Source Telecomputing Corporation
1616 Anderson Road
McLean, VA 22102
(800) 336-3366

TeleSports, Inc.
8547 Manderville Lane
Dallas, TX 75231
(214) 692-8787

Trinet, Inc.
9 Campus Drive
Parsippany, NJ 07054
(201) 267-3600, (800) 874-6381

Unlisted Market Service Corporation
49 Glen Head Road
Glen Head, NY 11545
(516) 759-1253, (800) 642-3841

U.S. Bureau of the Census
Center for International Research
Scuderi Building, Room 409
Washington, DC 20233
(301) 763-4286

VU/TEXT Information Services
1211 Chestnut Street
Philadelphia, PA 19107
(215) 665-3300, (800) 258-8080

Wilsonline
950 University Avenue
Bronx, NY 10452
(212) 588-8400

AUDIO BRIDGING COMPANIES

AT&T Communications—Alliance™ Teleconferencing Service
295 North Maple Avenue, Room 5242C2
Basking Ridge, NJ 07920
Contact: (201) 221-7138

American Teleconferencing Services, Ltd.
5503 Foxridge Drive
Mission, KS 66202
Contact: (800) 255-0390 or (913) 384-0066

	AT&T	Am. Telecon	Confertech	Conex Int.
1. Manufacturer of bridge	AT&T	Confer-Tech	Confer-Tech	Darome
2. Basic rates	25¢ per line per minute + LD charges	$7.50 to 15.00 per line per hour + LD	$20 per line per hour + LD	$20 per line per hour + LD
3. Discounts for quality usage	No	Yes	Yes	Yes
4. Total number of ports available	59	55	72	Not given
Call initiation options:				
5. Meet me	No	Yes	Yes	Yes
6. Operator dialed	Yes	Yes	Yes	Yes
7. Client dial each party	Yes	No	No	No
8. Fully automated	Yes	Yes	Yes	Yes
9. Dial-in 800 numbers	No	Yes	No	No, call collect available
10. Operator monitoring	No	Yes	Yes	Yes
11. Operator recall	Yes	Yes	Yes	Yes
12. Pre-call consulting at no charge	Yes	Yes	Yes	Yes
13. Call debriefing at no charge	No	Yes	Yes	Yes
14. Formal user training at no charge	Yes	Yes	Yes	Yes
15. Return audio for video conferences	Yes	Yes	Yes	Yes
16. Tape recordings made and on-line playback	No	Yes	Yes	Yes
17. Subconferencing	No	Yes	Yes	Yes
18. Customized answering upon request	Yes	Yes	Yes	Yes
19. Free on-line seminars	No	Fourth Thursday 10 A.M. CST	No	Fridays as scheduled 3 P.M. EST
20. Free bridge time	No	1 hour; up to 24 sites	1 meeting; no limit	30 minutes; no limit

Other services not included on the matrix which the companies would like noted:

AT&T Alliance™—None.

American Teleconferencing Services—Free speakerphone testing; free site testing (up to 30 minutes); audiographic support for slow-scan, facsimile (Group I and II), and other multipoint distribution transmissions.

ConferTech International—Data connections for slow-scan and other audiographic devices; international calling; listen-only mode is available on the bridge.

Connex International—WorldCall international service; Nightwire emergency service; IRS or instant reservation service; free equipment check-out; Private Label Service; Preferred Customer program.

Darome Connection—Networking of international conference sites; free customer seminars each week; conference equipment available for purchase or rental.

Multilink—Full duplex operation; computer display of conference for lines for each operator assistance.

Courtesy Virginia A. Ostendorf, Inc.

Darome Co. n.	Multilink	So. New England Tel.	Teleconn Nwk.	Teleconnect	Westell
Darome	Multi-link	CEAC and Mitchell	CEAC	Confer-Tech	Westell
$20 per line per hour + LD, ½ hour minute	23¢ to 28¢ per minute per line + LD	Meet-me: $6 + LD Auto ans.: $5 + LD Dial-out: Time of day oper. rate	$12 to 14 per line per hour + LD	Discount LD charges only	20¢ per minute per line + LD
Yes, but qualified	Yes	No	Yes	Yes	Yes
More than 500	100	Meet me: 17; Dial-out: 60	19	72	168
Yes	Yes	Yes	Yes	Yes	Yes
Yes	Yes	Yes	Yes	Yes	Yes
No	Yes	No	No	No	Yes
Yes	Yes	Yes	Yes	Yes	Yes
Yes	Yes	No	No	Yes, special discount	No
Yes	Optional	Optional	Optional	Yes	Optional
Yes	Yes	Yes	Yes	Yes	Yes
Yes	Yes	Yes	Yes	Yes	Yes
Yes	Yes	Yes	Yes	Yes	Yes
Yes	Yes	Yes, also has pay course	Yes	Yes	Yes
Yes	Yes	No	Yes	Yes	Yes
Yes	Yes	No	Yes	Yes	Yes
Yes	Yes	No	Yes	Yes	Yes
Yes	Yes	Yes	Yes	Yes	Yes
Every Wednesday 10:30 CST	On request	No	Yes as scheduled	No	Monthly as scheduled Fridays 1:00 CST
30 minutes; up to 10 sites	1 meeting; no limit	No	1 meeting; no limit	20 minutes; up to 6 sites	30 minutes; up to 5 sites

Southern New England Telephone Company—24-hour service, slow-scan and audio-graphic conferences (enhanced dialogue); international dialogue; demand appointments and profile reservations; customer-specific contacts.

Teleconference Network—Duplex operation, no voice block, easier moderator control; auto and manual volume control; custom programs designed telephone focus groups, customer seminars, etc.; specialized moderators provided; many participant invitation options.

Teleconnect—24-hour service, 7 days a week, low weekend rates; no premium charges; special low-cost 800 service package available; tape transcription at $5 per page.

Westell—Fully automatic system: Customer is issued "PIN" number, dials into system and controls dial-up, security and meet-me conference; detailed printed billing of all ports and numbers dialed is provided.

ConferTech International, Inc.
- 240 Applewood Tech Center, 2801 Youngfield
  Golden, CO 80401
  Contact: (800) 525-8244 or (303) 237-5151
- 742-D Hampshire Road
  Westlake Village, CA 91361
  Contact: (800) 525-8244 or (818) 889-3653
- 600 Maryland Avenue SW, Suite 220
  Washington, DC 20024
  Contact: (800) 525-8244 or (202) 863-0384

Connex International
- 12 West Street
  Danbury, CT 06810
  Contact: (800) 234-9430 or (203) 797-9060

Darome Connection
- 5725 N. East River Road, Suite 780
  Chicago, IL 60631
  Contact: (800) 435-6174 or (312) 399-1613
- 195 Main Street
  Danbury, CT 06810
  Contact: (800) 243-0991 or (203) 797-1300
- William House, 2nd Floor Worple Road
  Wimbledon, England SW19 4BA
  Contact: 011-441-879-1000
- 480 Second Street, Suite 302
  San Francisco, CA 94107
  Contact: (415) 896-1610
- 12400 Princeton Avenue
  Savage, MN 55378
  Contact: (612) 894-7780
- 90 Gerrard Street West
  Toronto, Ontario M5G 1J6
  Contact: (800) 268-9072 or (416) 585-9644
- 2100 M Street NW, Suite 606
  Washington, DC 20036
  Contact: (800) 835-9700 or (202) 331-3900
- 26 Broadway, Suite 1607
  New York, NY 10004
  Contact: (800) 922-1124 or 212-968-0990

Southern New England Telephone Company
- 310 Orange Street, Room 303
  New Haven, CT 06510
  Contact: (203) 497-6378

Teleconferencing Network
- 2 Prel Plaza
  Orangeburg, NY 10962
  Contact: (800) TEL-MEET or (914) 365-0123

Teleconnect
- 185 50th Avenue S.W.
  Cedar Rapids, IA 52404
  Contact: (800) 732-2487 or (319) 366-6600

Westell, Inc.
- 7630 Quincy Street
  Willowbrook, IL 60521
  Contact: (800) 323-6883 or (312) 325-5628

- 103 South Kenwood
  Glendale, CA 91205
  Contact: (818) 241-7699

- 5510 Columbia Pike, Suite 305
  Arlington, VA 22204
  Contact: (703) 671-5755

For information on international sites, call Chicago office.

Bridge Locations by Service

AT&T Communications Alliance™
White Plains, New York   Dallas, Texas
Chicago, Illinois   Los Angeles, California

American Teleconferencing Service
Kansas City, Kansas

ConferTech International
Golden, Colorado   Washington, D.C.
Los Angeles, California

Connex International
Danbury, Connecticut

Darome Connection
Chicago, Illinois   Danbury, Connecticut
London, England   San Francisco, California
New York, New York   Toronto, Ontario
Washington, D.C.

Multilink, Incorporated
Boston, Massachusetts

Southern New England Telephone Company
New Haven, Connecticut

Teleconference Network
Orangeburg, New York

Teleconnect
Cedar Rapids, Iowa

Westell, Inc.

Chicago, Illinois

Washington, D.C.

Madrid, Spain

Los Angeles, California

Paris, France

London, England

TELECOMMUTING BUSINESS CASE

To implement telecommuting, the following steps are necessary:

- Establishing corporate objectives.
  - Lay the groundwork with upper management.
  - Build the foundation with the telecommuters.
- Determining real versus perceived benefits.
  - Increased communications.
  - Shorter decision times.
  - Increased productivity.
  - Telecommuter needs.
  - Lowered travel and management "down time."
- Stages of implementation.
  - Check vendor lead times.
  - Check home facility availability.
- Analyzing user requirements.
  - User needs are primary.
  - Analysis is essential to meet those needs.
  - Aids in finding potential users.
  - Real need by telecommuters determined.
  - Who will do analysis—in house or consultant.
- Communications audit. A user audit would include analysis of the frequency and nature of the following:
  - Modes of communication:
    Voice.
    Data.
    Video-graphics.
    Text.
  - The internal office communication network to:
    Home.
    Neighborhood work centers.
    External, customers, audience, or membership.
    Other business and professional connections.

- The information network:
    Access to information.
    Storage of information.
    Transmission of information.
    Information-float waiting times.
    Load carried by the communications network.
        Home network may need to be replaced.
- Technology issues:
    Present office tools, yellow pads, calculators.
    Smart office tools, personal computers, teleterminals, etc.
    Office support systems, electronic mail, voice storage, dictation, micrographics, office supplies.
- The life support system.
  - Comfort heating, cooling, lighting, humidity.
  - Ergonomics—sitting, standing, moving about.
  - Psychological—visual, convenience, flexibility.
  - Personal—parking, exercise, day care, windows, dining, social.
- The work and meeting matrix.
  - Time required to plan.
  - Types, numbers, and relative importance.
  - Need for and use of graphics.
  - Nature of communication—real versus nonreal time.
  - Desire for flexibility in communication.
- The travel matrix.
  - Frequency—home-to-office.
  - Destinations, distances.
  - Exact departure and return times.
  - Attitudes towards travel—motivation for telecommuting.
  - Opportunity, costs of working with people at a distance.
- Cost analysis.
  - Telephone expenses—accountability.
  - Mailing costs.
  - Travel expenses, including actual (not only budgeted) expenditures for using automobiles, public transportation, parking, telephone tolls.
  - Review of all the other costs from life support systems to software systems for the worker.
- Analysis process.
  - Determine potential user groups.
  - Develop methodology based on those group activities.
  - Topics for investigation are:
      Travel information.
      Meeting information.
      Familiarity with telecommuting.
      Interest.

- Learn how user groups do business.
- Understand their needs.
- Describe benefits.
- Suggest uses tailored to them.
- Query their perceived needs.
- Don't expect complete understanding.
- Expect resistance.
- Have patience.
- Conduct extensive interviews.
- Analyze results.
- Economic evaluation.
  - Prepare introduction.
  - Outline business case issues.
  - Provide financial analysis summary.
    Capital expenses.
    Implementation costs.
    Recurring costs.
    Displaced costs.
    Comparison of alternative cases.
    Sensitivity analysis.
  - Present system considerations.
    Obsolescence.
    Compatibility.
    Room design.
    Service and support costs.
    Training required by staff.
    Backup and alternatives.
    System security.
  - Develop and present audit trail criteria.
  - Develop management and organization impacts.
  - Present telecommuting team and responsibilities.
- Introducing the system.
  - Begin the real work.
    Early awareness should take place now.
    Bring in more help, if needed.
    Develop series of user manuals.
      Pocket guide.
      Technical manual.
      Sales brochure.
  - Training
    Essential part of the system.
    Must be geared to users.
    Must hold their attention.
    Speak to training department for help.
    Keep it short, sweet, and graphic.
    Use videotape.

- Follow-up.
    Helps to ensure repeat usage.
    Realize it takes four or five months to understand benefits.
    Real measure of success is repeat usage.
    Observe other telecommuting environments if possible.
    Hold users' hands and guide them.
    Keep it up until they are convinced.
    Remember that satisfied users are the best advertising.
    Ask for candid reactions.
- Expansion plans.
    - Modify system according to telecommuters' suggestions.
    - Look for new users, applications, high communications usage.
    - Continue advertising.
    - Write system up for internal company organs and trade press.
    - Monitor usage, comments, trends, technology.

The business case is an important planning tool. It gives you a benchmark from which to develop new systems or to expand existing ones.

MAJOR PRODUCTS/SERVICES SUGGESTED FOR INTELLIGENT BUILDINGS

The services in which telecommuting could play a role are highlighted by an asterisk "*":

Intelligent Building Management

- On-line administration and financing.
- Computer-aided design space management.
- On-line facilities management and planning.

Intelligent Command and Control

- Life support and comfort—personal comfort environments.
- On-line power management.
    Uninterruptable power supply systems.
    Power generation.
- Wire management—building and local area networks (BAN/LAN).
    Power.
    Computer.
    Telephone.

- On-line building control.
    Remote administration.*
    Simulations.
- Maintenance.
    Repair and diagnostics.
- Security.
    Access and reporting.
    Protection.
    Remote monitoring.*
- Traffic.
    Flow.
    Accounting and control.

Information Technologies

- Multitenant—telecommunications.
    Video teleconferencing.*
    Audio bridging services.*
    Audio-graphic presentation equipment.
        Computer-generated slides.
    Sale, rent, lease of telephone station equipment.
    Sale, rent, lease of data communications systems.
    Sale, rent, lease of computer and office systems.
    Intercom-only calling service.
    Access (remote) to local calling service.*
    Standard and advanced calling features.
    Automatic—least cost—calling.
    Installation of systems.
    Maintenance of systems.*
    Moves and changes.
    Billing and management reports.
    Special account codes.
    Authorization codes.
    Toll restrictions.
- Messaging.*
    Telephone answering.
    Electronic text messaging and network interface.
    Voice mail.
- Network interface—carrier and bypass.
    Access to CATV networkings/programming.
    Access to external on-line databases.
    Remote access ports.*
    Modem pooling.
    Packet network interface.

- Paging.
    Speaker.
    Pocket pager.*
    Cellular telephone.*
- Tenant/text—building directory and information systems.
    On-line concierge.
- Archival storage.
    On-line storage.
    Vault storage and document shredding.
- Intelligent/movable/programmable office environments.
- On-line—word/data/info processing.
- Optical character readers and facsimile/telex.
- Tenant support and training.
    Computer-aided instruction (CAI).
- Encryption.
    Systems security.

Intelligent Resources

- Conference and situation "war" rooms.
    On-line room scheduling.
- Temporary services.
    Leased staff.*
- Package, courier, and delivery services.
- Travel, conference, meeting planning services.
- Day care and health/recreational gyms.
- Print/copier/office supplies services.
- Interior design services.
    Computer-aided design.
- Educational/training services.*

THE GRAPEFRUIT DIET

Implementing Electronic Communications

Implementing any new technology requires an enormous amount of time and energy. The following steps are suggested for the organization that integrates a telecommuting program.

1. *Start now*—Too many organizations "play around" before making the decision to start. If plans are kept simple and pilot projects move along, progress will be made, albeit slowly. Most importantly, take the new technology in "small bites." Organizations should consider one system, apply it, modify it, learn from it, and expand it. Also, make sure there is a leader to pilot the project. Without this effort, a program is likely to fail.

Telecommuting may be difficult, if not impossible, in some situations. However, for people who must work at home, it can be used effectively and will provide the best technology.

2. *Think small*—Take small steps. Find pockets of people who are interested in getting started. Work with them for a while, then find another group, using the first as a model.

3. *Let the monster sleep*—If you have employees who don't want to telecommute, ignore them. These people may change their attitudes, but it is more likely that, over time, technology will change to be more compatible with them. Telecommuting is not for everyone, nor useful all the time. Allow employees to understand what the technology will do for them. And remember that cars, television, and airplanes are still not accepted by some people.

4. *Understand the enemy*—The difficulty in adapting to new technology often resides in the tasks that are performed on it. It is plain to see that nearly everyone has a different job title and manages the specialized information that goes with it. Robotics and factory automation have been successfully incorporated into the manufacturing environment because by programming machinery to assume repetitive tasks, workers are freed to pursue more interesting undertakings.

   As we move into upper-level management positions, jobs become vastly more complex and more dependent on individual management styles. Here, automation in general and telecommuting in particular can hope to be an "executive helper," at best. It supplements travel rather than replaces it. It should also be viewed as a "power tool" that helps managers gain visibility, handle workloads, and increase work quality.

5. *Sell the sizzle*—Marketing is the key to office automation and will help promote telecommuting. A term coined years ago, "organizational marketing," describes the tactics and strategies that can be used to sell programs within an organization. Motivational techniques, posters, ads, promotional devices, and (even) cash incentives are sometimes needed to encourage program participation.

6. *Train the beast*—Training will be a limiting factor until management has mastered its telecommuting skills. Like most other technologies, telecommuting suffers from an enormous failure rate. People think telecommuting is going to be great, but when they fail at first, they give up.

7. *Persist*—When an organization has a vision and the means to implement it, there is nothing in the world that can take the place of persistence.

Appendix A

Audio Bridging Features

The primary issues in selecting an audio bridge include:

- Speech quality.
- Ability to interrupt.
- Number of concurrent conferences.
- Size of conference.
- Ease of set-up speed.
- Ease of operator training.
- Security.
- Graphic terminal compatibility.
- Transmission of data communications or slow-scan.
- Unattended operator capability.
- Technology—analog or digital.

Operational features of most audio bridges include:

- Forty-eight port teleconference capacity.
- Microprocessor controlled with automatic answer and instructions by voice synthesizer.
- Access by off-premise tone dial keypad.
- Software allowing subdivision of teleconference into simultaneous subteleconferences.
- Automatic gain amplification compensating for network losses to provide high-quality audio to each participant.
- Security features via touch pad which lock out undesired participation.
- Software customized to user's operational requirements.
- Monitor jacks that provide "operator" interface, if desired.

- Individual status lights on each port module to allow talker identification.
- Expander jack that allows linking of two systems for full 48-port configuration.
- Twenty-four hours a day access from any telephone in the world network.
- Compatibility with all types of amplified speakerphones or teleconference systems.

Characteristics of audio teleconferencing include:

- Greetings, introductions, and farewells. These must be incorporated into the conference because arrivals and departures are invisible.
- Participants can remain relatively anonymous.
- Active listening is required if conferees are to be involved in a meaningful way.
- Vocabulary and syntax have a heightened impact on listeners, hence on the conference effectiveness.
- Vocal characteristics such as inflection, volume, and speed magnify the impact of the communication exchange. Pauses in conversations and long silences will also contribute to its drama.
- Difficulty in sending or receiving messages that depend on such visual cues as eye movements or body language.

Preparation for an audio teleconference is vital to its success. Some of the key issues are:

- Installation and testing of equipment.
- Schedule of program dates.
- Deadlines set for handouts and audiovisual material(s).
- Instructor(s) contacted/trained regarding their role(s).
- Coordinators selected for each participating site.
- Guidelines given instructor(s); these include information on participants, site location, and names of coordinators.
- Coordinators given handout/training explaining their roles/duties.
- Practice run of program.

Appendix B

Teletraining as a Telecommuting Communications Tool: Lessons Learned from Teletraining

Interview with Virginia A. Ostendorf—Virginia A. Ostendorf, Inc., October 14, 1985.

Having worked in teletraining for the past six years, Virginia Ostendorf formed many opinions and reached many conclusions regarding her work. Those she considered the most important are given below:

1. No one form of teleconferencing is superior to any other for the delivery of teletraining.

Nonusers universally assume that full-motion video teleconferencing is essential for an effective teletraining program. Ostendorf noted, "Although full motion video teleconferencing *is* a wonderful training medium, audio and audio-graphic systems can work equally well for most projects. It is wise to limit preconceived notions and work with what is affordable and available."

2. Teletraining classes should not strive to be like face-to-face classes.

Those who attempt to duplicate face-to-face training in the electronic medium are doomed to failure. She states, "Teletraining is *not* traditional training. It is different. Enjoy and explore those differences. Find what teletraining allows you to do that traditional training does not. Only then will you begin to look at teletraining as something other than a second class citizen."

3. Teletraining is most cost effective when the number of persons to be trained is large.

Most trainers think of the ideal group as 20 to 25 persons. Teletraining enables you to reach multiple groups of 20 to 25. She suggests, "Try to set aside notions of training based on other systems. Instead, look at the training need, look at the content, and where the prospective students are located." The possibilities for on-line and off-line activities in this configuration are endless.

4. There is no optimum number of attendees for a teletraining class. There is no optimum length for a teletraining class. There is no optimum number of sites for a teletraining class.

Again, the rule books have not yet been written. But if Ostendorf were to write one, she would say, "break the rules." Most teletraining beginners are hesitant to risk a multipoint conference. Yet she has found multipoint conferences are far more exciting to teach. Teletraining classes can effectively train 5 persons or 500. Classes can last 30 minutes or 2 weeks. Naturally, to tackle a two-week effort requires considerable experience or professional help. But there are no hard and fast rules.

5. Teletraining calls for special techniques and presents unique problems.

Do not put a traditional trainer on a teleconference and expect success. She notes, "no matter how skilled or charismatic, the traditional trainer needs to learn planning and presentation techniques appropriate for the medium. Discussion techniques are crucial, particularly in nonvisual systems." Knowing how to handle special problems such as time zones, mike fright, and equipment difficulties is important to a polished presentation. The development of appropriate support materials is another topic of importance when training the traditional trainer.

6. Teletraining can be boring.

This is the major sin of most teletraining conducted by nonprofessionals. According to Ostendorf, "If techniques are not known, if support materials are weak, if pacing is poor, and if the lesson plan is inadequate, the student becomes bored." Most employees expect sophisticated presentations, whether in films, videotapes, television, or face-to-face productions. It is very important that planning for teletraining include variety, stimulation, and interaction. These are the weapons that combat boredom.

7. Trainer resistance to teletraining is common.

Few traditional trainers have any experience with teletraining. It is easy to see why initial resistance to the concept is commonplace. According to Ostendorf,

> Do provide opportunities for role modeling using experienced teletrainers. Let your trainers see teletraining in action. Do select your first in-house teletrainers on a voluntary basis. Telling traditional trainers that they must teach this way is not recommended. Instead, call upon those to whom the concept is intriguing, and provide them with ample support.

It is often wise to develop new courses for teletraining delivery rather than threaten the empire of a traditional trainer who would rather teach in the "old" way. New technology courses, such as computers, word processing, new telephone systems, and the like, are well suited to introduction via teletraining.

8. Many companies with a need for teletraining never use it.

Companies who now have successful teletraining programs at one time had none. Ostendorf noted, "These are the risk-takers who had a training need and established a pilot project using teletraining." Particularly where equipment can be leased and little capital expense is needed, corporate trainers should take a chance on teletraining. The risk is minimal; enough examples abound to convince even the most skeptical trainer that teletraining is realistic and effective.

One characteristic is common among trainers converted to teleconferencing. "They are enthusiastic," according to Ostendorf. In a department with little corporate influence, it is easy to lose heart. Teletraining offers the corporate trainer a chance to train more people, to develop more courses, to learn new skills. It offers management significant benefits, not the least of which is a better-trained work force. Experienced teletrainers find that this form of electronic communication is the most human, the most personal, and the most satisfying training mode of the many offered in the last decade.

According to Ostendorf, "When geography is no obstacle, training can come into its own, reaching even the most remote company location with pertinent, timely information crucial to the job. Teletraining will be at the heart of the revolution."

Appendix C ===

Voice Mail Features

The options of voice store and forward systems are narrowed to two specific applications:

1. Stand-alone applications such as IBM, Wang, VMX, and others provide.
2. PBX applications such as ROLM, Northern Telecom, Anaconda Ericsson, and others provide.

Listed below is a desired-feature list of voice store and forward systems and a comparison guide:

Feature Comparison	PBX Controlled	Stand-Alone
Telephone answering	Yes	Yes
Personalized greeting	Yes	Yes
Message notification	Yes	Yes
Listen to messages	Yes	Yes
Record messages	Yes	Yes
Send messages:		
Send	Yes	Yes
Automatic reply	Yes	Yes
Forward	Yes	Yes
Ringing delivery	—	Yes
Timed delivery	—	Yes
Manage messages:		
Stop/start	Yes	Yes
Replay	Yes	Yes
Skip	Yes	Yes
Edit	Yes	Yes

SOURCE: Cross Information Company.

Glossary

Abbreviated Addressing—Using a short code to send a message to a preprogrammed address.

Access—To key into a system, so storing or retrieving information is possible.

Access Time—The interval between the time that information is called from storage to the time delivery is completed.

Account Number—A number that identifies a budget/billing unit and associated storage areas.

Acoustic Coupler—A modem that cradles a telephone handset during transmission. A microphone in the modem picks up telephone tones and translates them into digital signals that can be understood by the receiving computer. Conversely, a modem translates digital signals from a computer into audible tones which are transmitted over telephone lines.

Action-Item List—A list that contains reminders that are brought to a member's notice periodically or on a specific date.

Active Terminals—Terminals that are prepared to receive documents.

Alphanumeric—Letters, numbers, and special characters.

Analog Signal—A continuous signal that varies in direct proportion to the strength of an input signal. Telephones transmit the human voice by converting sound waves into electrical analog signals.

ANSI—American National Standards Institute. A group whose codes and standards apply to the computer industry.

Answer/Auto-Answer—A modem function whereby the equipment senses an incoming ring signal on the telephone line and automatically connects the modem to the line.

Application—A specific program or task. Sorting employee records, which a computer can execute, would be an example.

Application Program—A computer program designed to meet specific user needs, such as controlling inventory or monitoring a manufacturing process.

Applications Software—A computer program designed for a specific use, as for bookkeeping.

Architecture—The design or organization of a system. Computer architecture refers to the central processing unit.

Archive—A record of information. One stores messages on archive disks.

Arithmetic Capability—Ability of the system to perform calculating functions, such as addition, subtraction, multiplication, and division.

Array—Data in a row or matrix arrangement.

Artificial Intelligence (AI)—Computer programming that recognizes ideas and answers problems. A system may have sensory perception. AI is used in robotics and "expert systems" where there is a base of specialized knowledge and a program from which a computer can solve problems.

ASCII (American Standard Code for Information Interchange) Code—A binary number assigned to each alphanumeric character and several nonprinting characters that are used to control printers and communication devices.

Assembly Language—A programming language based on a computer's structure and machine language.

Assistant—An assistant in a computer conference group or center is responsible for managing the public/private resources within his/her scope. Assistants are prohibited access to the content of items stored in private files. They serve as advisors within their organizational units.

Asynchronous—Not synchronized to work together. Asynchronous transmissions can be sent or received when participants choose, as opposed to being sent at fixed intervals.

Attention Stack—A storage facility for unfinished material (notes, messages, documents) that preserves their status while the user works on something else.

Audio—Voice portion of a communications link.

Audio Conferencing—Holding a telephone conference.

Auto Answer—A feature that allows equipment to automatically receive and store information or messages until the recipient requests them.

Auto Dial—A modem function that enables a modem to dial telephone numbers and establish computer connections.

Automatic Carrier Return—A system that automatically performs a carrier return. This permits the operator to continue typing without pausing at the end of each line.

Automatic Pagination—Division of a multipaged document into pages having a specified new number of lines per page.

Automatic Polling—A system whereby one computerized device automatically requests that another transmit information. Operators can specify times for equipment to "make calls," send, and collect messages.

Automatic Sorting and Listing—Sorting of items in numerical or alphabetical sequence, or topic, by which the file can be organized.

Background—Noninteractive services running on a computer while a person is using an interactive service.

Background Processing—The execution of a low-priority computer program when higher-priority programs are not using the system's sources.

Backup—Copies of one or more files on a storage medium for safekeeping.

Bandwidth—The difference between the highest and lowest frequencies that a transmission channel can carry. A standard telephone line generally carries frequencies between 300 and 3,000 hertz (cycles per second), providing a bandwidth of 2,700 hertz.

Baseband Transmission System—A means of transmitting data on a local area network using:
- Coaxial cable or twisted pair wire.
- Digital signals.
- Inexpensive transceivers.
- Distributed control.

BASIC (Beginners' All-Purpose Symbolic Instruction Code)—A widely used interactive programming language developed by Dartmouth College. It is especially well suited to personal computers and beginning users.

Batch Processing—A technique of executing a set of computer programs without human interaction or direction. Under certain conditions direct interaction is possible.

Baud—A unit for measuring data transmission rates. Technically, baud rates refer to the number of times the communications line changes state after each second.

Bell 103—A protocol standard developed by Bell Telephone for modem communication at speeds below 300 baud.

Bell 202—A protocol used by modems for half duplex transmission at speeds of up to 1,200 baud.

Bell 212—A protocol used by modems for full duplex transmission at speeds of up to 1,200 baud.

Bidirectional—(1) Ability to transfer data in either direction. This is characteristic of a "bus" local area network. (2) Ability of a print head to print from right to left and from left to right. This increases print speeds.

Binary—The fundamental number system used with computers. Binary numbers are represented by only two numerals, 0 and 1. The binary system is necessary because electrical circuits store and sense only two states: ON and OFF.

BISYNC (Binary Synchronous Communication)—A method of transmission normally used by IBM mainframes. BISYNC gathers together a number of message characters and puts them in a single large message block that includes special characters, synchronized bits, and station addressing information.

Bisynchronous—Binary and synchronous signaling.

Bit (Binary digIT)—A unit of information that designates one of two possible values. A bit is usually written as a 1 or 0 to represent the ON or OFF status of an electrical switch.

Bit-Map Graphics—A technology that allows control of individual pixels on a display screen and that produces graphic images of superior resolution. It permits accurate reproduction or arcs, circles, sine waves, or other curved images that block-addressing technology cannot accurately display.

Black Box—A device that connects incompatible hardware or software so that it can interact.

Board—Also "circuit board." A plastic resin board containing electronic components such as chips and the electronic circuits needed to connect them. See option module.

Boot—To start up a system or program. A cold boot means a first start.

BPS—Bits Per Second.

Bridge—An electronic "place" where three or more people can confer by telephone, or in some cases by data terminal.

Broadband Bus Network—A coaxial cable network that is divided into a number of high capacity channels able to carry data traffic, video, and voice transmissions. Many bits per unit of time can be moved from point to point.

Broadband Channel—A communications channel with a large bandwidth or capacity. Channels wider than voice grade are often considered to be broadband.

Broadcasting—Sending the same message to a large number of people at the same time, from point to multipoint.

Bubble Memory—A type of memory composed of small magnetic domains formed on a thin crystal film of synthetic garnet. This system operates rapidly and holds informational content when power is turned off.

Buffer—A temporary memory storage area for information until equipment is able to process it. A buffer also holds data being passed between computers or other devices such as printers which operate at different speeds or different times.

Bug—Error in hardware or software programming.

Bulletin Board—An electronic file within an electronic mail or teleconferencing system. All participants can place or access public messages placed there.

Bus—A group of parallel electrical connections that carry signals between computer components or devices within a local area network.

Byte—The number of bits used to represent a character. For personal computers, a byte is usually eight bits.

Cable—A group of conductive elements, such as metal wires or fiber optic cable, packaged as a single line to interconnect communications systems.

Cable Television—A telecommunication system that uses coaxial cable to distribute the TV signal.

CAI—Computer-Aided Instruction.

Calendar—A calendar is a permanent storage area associated with a group. It contains information about appointments, business trips, meetings, vacation times, etc., for each member of a conference group.

Carrier—A company that provides transmission capabilities for the general public.

Cathode Ray Tube (CRT)—A vacuum tube that generates and guides electrons onto a fluorescent screen to produce characters or graphic displays on video display screens.

CATV (Community Antenna Television System)—Coaxial cable system that transmits television or other signals to subscribers from a single head-end location.

CBMS (Computer-Based Message System)—A sophisticated computer system that receives, stores, and transmits messages. Messages are delivered to the electronic mailboxes assigned to each user.

CCITT—Consultative Committee for International Telephony and Telegraphy, a part of the International Telecommunications union that sets communications standards for the industry.

Center—A center is a teleconferencing unit that consists of people who have common interests or tasks and are provided with certain facilities. Its purpose is to identify a number of people with a common interest or task and provide them with certain facilities like a portfolio, bulletin board, and calendar to support their cooperation.

Center Manager—A center manager establishes new centers, enrolls new members, and assigns members roles, e.g., group manager or center assistant.

Central Processing Unit (CPU)—Electronic components that cause work in a computer to occur by interpreting instructions, performing calculations, moving data in main computer storage, and controlling the input/output operations. A CPU consists of the arithmetic/logic unit and the control unit.

Channel—A band of frequencies allocated for communications.

Character—A single printable letter (A–Z), numeral (0–9), or symbol (,%$.) used to represent data. Text symbols such as a space, tab, or carriage return are not visible as characters.

Character Code—Numerical values assigned to characters. The ASCII code is an example.

Character Printer—Equipment that prints one character at a time like a type-writer.

Character Set—The characters of a code, font, or device that can be generated and displayed.

Chip—Semiconductor material containing microscopic integrated circuits.

Circuit—(1) A system of semiconductors and related elements through which electrical current flows. (2) In data communications, the electrical path providing one-way or two-way communication between two points.

Circuit Switching—The physical connection taking place between channels.

Coaxial Cable—Cable that has one insulated conducting wire at the center. A second wire surrounds the insulation and is also insulated. It supports large bandwidth, has high data rates, high immunity to electrical interference, and low incidence of error.

COBOL (Common Business-Oriented Language)—A high-level programming language, developed for the U.S. Army, that is well suited to business applications involving complex data records (such as personnel files or customer accounts) and large amounts of printed output.

CODEC (COder/DECoder)—A chip in a telephone that converts an analog signal to a digital pulse, or the reverse.

COM (Computer Output Microfilm Equipment)—These devices are used to record computer output as very small images on a roll or sheet of film.

Command—A user's instruction to the computer, generally given through a keyboard. This can be a word, mnemonic, or character that causes a computer to perform an operation.

Communicating Word Processor—A specialized computer equipped to send and receive messages.

Communication Capability—The ability to transfer information between systems on and off premises.

Communications Interface—A system for converting keyboard signals to signals a network will accept.

Communications Protocol—See Protocol.

Communications Satellite—A satellite used to receive and retransmit data, including video and audio signals.

Compatibility—(1) The potential of an instruction, program, or component to be used on more than one computer. (2) The ability of computers to work with other computers that are not necessarily similar in design or capabilities.

Compiler—A translator that renders a program into a computer's own machine language.

Computer—A programmable machine made up of a microprocessor, memory, keyboard, monitor. It operates as a unit when instructions and power are applied.

Computer Architecture—Internal computer design based on the types of programs that will run on it and the number that can be run at one time.

Computer Network—An interconnection of computer systems, terminals, and communications facilities.

Computer Program—Instructions that tell a computer to do a specific task.

Computer Teleconferencing—Interactive group communication in which a computer is used to receive, hold, and distribute messages between participants for many-to-many communication.

Concentrator—A device that joins several communication channels together.

Configuration—The arrangement of equipment (disks, diskettes, terminals, printers, etc.) in a particular system.

Core—The older type of nonvolatile computer memory made of ferrite rings that represent binary data by switching the direction of polarity of magnetic cores. Most modern computers use integrated circuits which are faster than core memory, but are volatile.

CPU—Central Processing Unit.

Cross Training—Training employees in each other's duties.

CRT—See Cathode Ray Tube.

CSMA/CD—Carrier Sense Multiple Access with Collision Detection. A system by which equipment on a LAN seeks to transmit a message. When it senses the network is idle, it will send information.

Cursor—A movable, blinking marker—usually a box or a line—on the terminal video screen that indicates the next point of character entry or change.

Cursor Position—The place where a specific function should be performed. "Home" is the upper left corner, and "reverse home" is the lower right corner.

Daisy Wheel Printer—A printer that has letters and numbers on spokes that radiate from a plastic or metal wheel. As the wheel spins across the page, a hammer hits the spokes, pressing the characters against the ribbon and paper.

Data—Facts, numbers, letters, and symbols that can be stored in a computer. For personal computer users, data can be thought of as the basic elements of information created or processed by an application program.

Data Bank—A collection of data which is stored on auxiliary storage devices.

Data Code—A binary representation of a letter or number used by particular equipment.

Data Communication—The movement of coded data from a sender to an addressee by means of electrically transmitted signals on telephone lines, coaxial cables, microwaves, or by other means.

Data Diskette—A diskette that is used entirely or primarily to contain data files.

Data Processing—The application in which a computer works primarily with numerical data, as opposed to text. Many computers can perform data and word processing.

Data Set—(1) Another name for a modem. (2) A group of data elements.

Database—A large electronic collection of organized data that is required for performing a task. Typical examples are personnel files or stock quotations.

Database Management Software—A system of integrated tools to store, retrieve, and maintain a large collection of data. Some of the tools support functions like batch reporting, interactive query, and decision support.

Database Management System—(1) A series of programs used to establish, update, and query a set of facts, figures, or any other information, e.g., reservation system. (2) Software that controls the organization and access to database information.

Decentralized Processing—An arrangement whereby computers at remote locations communicate with a central processing unit but not directly with each other.

Dedicated Computer—A computer used for one special function, such as controlling the Space Shuttle's navigation system.

Degradation—Slowing of a data transmission as more users access a computer network or for other reasons.

Delete Capability—The method(s) a system uses to delete information. It takes place by removing a document, page, paragraph, word, or character string.

Device—In computers, it is hardware that performs a specific function. Input devices (e.g., keyboard) are used to enter data into the CPU. Output devices (e.g., printer or display monitor) are used to take data out of a computer in some usable form. Input/output devices (e.g., terminal or disk drive) are able to perform both activities.

Diagnostic Program—A program that checks the operation of a device, board, or other component for malfunctions and errors, and reports its findings.

Digital Signal—A series of electrical impulses that carry information in computer circuits.

Digitize—To translate voice or pictorial signals into binary code (digital format) for transmission.

Direct Connect Modem—A modem that plugs directly into a telephone outlet, bypassing the handset. It enables users to send and receive signals directly to and from telephone lines. See Acoustic Coupler.

Direct Distance Dialing (DDD)—Accessing telephones tied to the public switched network by using an area code.

Direct Memory Access (DMA)—A method for transferring data to or from a computer's memory without CPU intervention.

Directory—An index used by a control program to locate blocks of data that are stored in separate areas of a data set in direct access storage.

Disk—A flat, circular plate with a magnetic coating for storing data. Physical size and storage capacity of disks can vary. There are hard disks, optical disks, and diskettes, also called floppy disks.

Disk/Diskette Drive—A unit used to read data from or write data onto one or more diskettes.

Diskette—A flexible, flat, circular plate that is permanently housed in a black paper envelope. It stores data and software on its magnetic coating. Standard sizes vary in diameter. Diskettes are often called floppy disks.

Display Screen—A device that provides a visual representation of data; a TV-like screen often called a monitor, cathode ray tube (CRT), or video display unit (VDU).

Distributed Data Processing—A computing approach in which an organization uses a number of computers located at a distance but connected to an office computer.

Distributed Intelligence—An arrangement whereby terminals and peripheral equipment in a system possess a certain amount of intelligence and do some work, eliminating part of the main computer's burden.

Distributed Processing—A network system where work is distributed among connected computers and processed by them.

Distribution List—A list that identifies a collection of members. The name of the distribution list serves as shorthand in addressing groups collectively.

Documentation—The training manual that explains a program.

Dot-Matrix Printer—A printer that forms characters from a two-dimensional array of dots. More dots in a given space produce characters that are more legible.

Double Density—A special recording method for diskettes that allows them to store twice as much data as normal, single-density recordings.

Down Loading—Transferring a file or program from a central computer to another computer.

Downtime—The period of time when a device is not operating.

Draft-Quality Printer—A printer, usually high-speed dot matrix, that produces characters that are very readable but of less than typewriter quality. They are typically used for printing internal documents where type quality is not a major factor.

Drive—A peripheral device that holds a disk or diskette so that the computer can read data from and write data onto it.

Dumb Terminal—A terminal that consists of a keyboard and an output device such as a screen. A dumb terminal is used for simple input/output operations and generally has no intelligence of its own.

Duplex—See Full Duplex.

EBCDIC (Extended Binary-Coded Decimal Interchange Code)—A standard communications code consisting of an eight-bit coded character set. This code is used primarily by IBM mainframe computers.

Electronic Blackboard—The generic name for audiographic devices used to send writing over a normal telephone line. As the sender writes on a board, the writing apears at the distant location on a television monitor.

Electronic Circuit—A pathway or channel through which electricity flows.

Electronic File Cabinet—An electronic storage unit that files data in much the same way as a regular file cabinet. It has some distinct advantages: a great deal of information can be stored in a small area, accessed and changed quickly, organized more efficiently, and kept more securely.

Electronic Handshake—An arrangement whereby devices that transmit data can query receiving equipment regarding transmitting speeds, mode selection, line quality, and other conditions for the most compatible transmitting conditions.

Electronic Industries Association (EIA)—A standards organization located in Washington, D.C., that specializes in the electrical and functional characteristics of interface equipment.

Electronic Mail—A system that allows memos or messages to be sent from one or more person(s) or electronic device(s) to others.

Emulator—A program that allows a computer to imitate a different system. This enables different systems to use the same data and programs to achieve the same results but at possibly different performance rates.

End-to-End Connection—A through and open channel, like a telephone connection when people are speaking.

Ergonomics—The science of human-machine interaction.

Error Message—Text displayed by the computer when an incorrect response is typed. It explains the problem and indicates what to do next.

Expert System—Advanced computer programming that relies on a large body of specialized knowledge to give information on a professional task. Expert systems are also known as knowledge-based systems and are used for high-level management or complex applications.

Facsimile (Fax)—A process of scanning text or graphic material whereby the image is converted to signals. The signals are transmitted by telephone to a compatible terminal which is able to produce a copy of the original material.

Fanfold Paper—A continuous sheet of paper folded accordian-style and separated by perforations. It is used for computer printouts.

Field—The smallest unit of information or data within a record.

File—A collection of logically related records or data. A file is the means by which data is stored on a disk or diskette so it can be used at a later time.

File Organization—A system that determines the physical placement of data on a mass storage device.

Filename—The sequence of alphanumeric characters assigned by a user to a file so it can be read by the computer and the user.

Firmware—Software placed permanently on a "Read-Only Memory" (ROM) chip within a computer. It cannot be lost if power goes down.

Flaming—Anger that is produced by an electronic message.

Floppy Disk—A flexible magnetic disk that looks like a small phonograph record and is used for information storage. Such disks can be erased and reused. They are also called diskettes.

Flowchart Symbols—Standard symbols used to diagram programming logic.

Font—A device that prints or reproduces a specific typeface. See Typeface.

Foreground Processing—Top-priority processing; it has priority over background (lower-priority) processing.

Form Definition Software—A software package that provides facilities to design a screen display, including field protection and data verification.

Formfeed—A printer feature that automatically advances a roll of fanfold paper to the top of the next page or form when the printer has finished printing one page or form.

FORTRAN (FORmula TRANslation)—A widely used high-level programming language well suited to problems that can be expressed in terms of algebraic formulas. It is generally used in scientific applications.

Freeze-Frame Transmission—Transmission of high-quality, still motion images, about one each 35 seconds. It is also called slow-scan or still-frame teleconferencing.

Frequency Division Multiplexing—A modulation technique that divides the total capacity of a channel into specific frequency bands.

Full Duplex—A method of communication between two computers that allows transmission in two directions at a time.

Full-Motion Video—Continuous motion television images that provide interactive group communications.

Function Key—A key that causes a computer to perform a function (such as clearing the screen) or execute a program. On some personal computers, some function keys, such as HELP and DO, and all the arrow keys have predefined actions.

GIGO ("Garbage in; garbage out")—Refers to the fact that processing bad input data will result in bad output.

Gateway—A special node that interfaces two or more dissimilar networks and provides protocol translation between them.

Global—Refers to an operation encompassing a complete area, like a file, program, or database.

Graphics—The use of lines and figures to display data, as opposed to the use of printed characters.

Group—A group is an organizational unit within a computer conference.

Half-Duplex (HDX)—A network where data can be transmitted in both directions, but only in one direction at a time. Speakerphones are an example of a half-duplex system where one pushes a button before speaking or to interrupt another person.

Handshaking—An exchange of predetermined signals between two computers or between a computer and a peripheral device. It allows the computer to ascertain whether another device is present and ready to transmit or receive data.

Hard Copy—Output in a permanent form, usually on paper or paper tape.

Hard Disk—A disk such as a Winchester disk that is not flexible. It is more expensive than a diskette but capable of storing much more data.

Hardware—The physical equipment that makes up a computer system and permits information storage and transmission.

Hardware Interfaces—The plugs and cables that connect equipment components.

Hardwired—A permanent physical connection between two points in an electrical circuit or between two devices linked by a communication line. Personal computer local network connections are typically hardwired.

Hash Key—A key that acts like an address.

Head—A component of a disk drive that reads, writes, or erases data on a storage medium such as a diskette or disk.

Help Service—Information displayed on the video screen that explains how to use applications and system services.

Hertz—A unit of frequency equaling one cycle per second.

Horizontal Scrolling—Horizontal movement of text to access more characters than are shown on the screen.

Host Computer—The controlling computer in a multiple computer operation.

Icon—Graphic image often used in place of words. For example, a scissors may appear on a monitor next to a word that is being cut out of the text.

Impact Printer—A printer that forms characters on paper by striking an inked ribbon with a character-forming element.

Inbox—A file for incoming electronic mail.

Incompatible Devices (Equipment)—Equipment that doesn't interact effectively (communicate) with another device.

Index—An electronic table of contents stored in the computer to aid in the search for material on storage media.

Information Management—Evaluation and modeling tools that use information stored in a well-structured data system.

Information Services—Publicly accessible computer repositories for data, such as stock exchange prices, or foreign currency exchange rates and other databases.

Instruction—A command that tells the computer what operation to perform next.

Integrated Circuit (IC)—A complete electrical circuit on a single chip.

Intelligent Terminal—A terminal that is capable of processing information; many store and retrieve information on their own tapes, disks, and printers. An intelligent terminal can be adapted to communicate with various host computers simply by changing the protocol programmed into it.

Interactive Computer—Equipment that is capable of carrying on a dialogue with the user via a keyboard.

Interactive Software Package—A program that provides the user with commands with which to submit his requests and exercise control over the execution of the program.

Interface—A hardware connection that provides an electronic pathway for signals, or software that enables information to be exchanged between programs. Keyboards interface people and processors.

Internal Memory Capacity—Maximum number of characters that internal memory of a system can hold.

I/O (Input/Output) Devices—Equipment that works with a processor. Information may be entered or extracted by using them. A keyboard and display screen are examples.

Inverse Video—A reversal of foreground and background on a terminal display screen. White characters would be shown on black instead of the reverse.

IRC (International Record Carrier)—These are worldwide communications networks like RCA Global Communications and ITT World Communications.

ISO—International Standards Organization.

ITU—International Telecommunication Union, located in Geneva, Switzerland.

Job—A computer task (program), such as reading a disk or printing a file.

K—The symbol for the quantity 2 to the 10th power or 1,024. The K is uppercase to distinguish it from a lowercase k, which is a Standard International Unit for "kilo," or 1,000.

Key System—A small internal company telephone system without switching capabilities.

Keyboard—Typewriter-like terminal keys used for data entry.

Keyword—A word used to characterize the content of a file, document, or message. They are used in indexes of a file.

Kilobyte (Kb)—1,024 bytes.

Knowledge-Based Management System—A system that searches for, organizes, controls, increases, and updates an area of knowledge. KBMSs are part of expert systems.

Knowledge Engineering—Designing knowledge-based programs and expert systems.

LAN—See Local Area Network.

Large-Scale Integration (LSI)—The combination of about 1,000 to 10,000 circuits on a single chip. Typical examples of LSI circuits are memory chips, microprocessors, calculator chips, and watch chips.

Leased Line—A permanent dedicated point-to-point or multipoint telephone circuit used for transmitting voice or data signals. The line is leased from a long distance telephone company (common carrier) such as AT&T, and can be conditioned to permit higher transmission speeds than a standard line (see Voice Grade Line).

LED (Light Emitting Diode)—A semiconductor diode which emits light when it is charged with electricity.

Letter-Quality Printer—A printer used to produce final copies of documents. It produces typing comparable to that of a high-quality office typewriter.

Light Pen—A device that allows data to be entered or altered on a CRT screen.

Line Speed—See Data Communications.

Lineprinter—A high-speed printer that produces an entire line of characters at one time.

List Processing—The word processing application that permits many copies of a form document to be produced, with certain information changing from one copy to the next (e.g., the production of personalized form letters).

Local—Hardwire connection of one computer to another computer, terminal, or peripheral device such as in a local area network.

Local Area Network—A communications network connecting computer terminals and other devices within an organization. LANs may also connect with other private or public networks.

LSI—See Large-Scale Integration.

Machine Language—A system that defines instructions that a computer can carry out.

Magnetic Tape (Magtape)—Tape used as a mass storage media, and packaged on reels. Since the data stored on magnetic tape can only be accessed serially, it is not practical for use with personal computers. It is often used as a backup device on larger computer systems.

Mail Database—A depository of currently pending mail.

Mail Qualifier—An attribute of information. Examples are:
1. Recipient.
2. Sender.
3. Forwarding permission.
4. Copy permission.
5. Special mailcode.
6. Request for response.

7. Site ID.
8. Modification capability.
9. Keep capability.

Mailname—User identification which is unique within a group, possibly the teleconferencing center, or even the system.

Mainframe—Centralized computer facility (CPU and main memory). It may delegate some of its workload to specialized processors.

Mass Storage—A device such as a disk or magtape that can store large amounts of data readily accessible to the central processing unit.

Matrix Management—An arrangement where work is organized around organizational work groups.

Mbyte (MB)—1,048,576 bytes, or 1 million.

Medium—The physical substance upon which data is recorded. Magnetic disks, magnetic tape, or punched cards are examples.

Meet-Me Bridge—A dial-up audio conferencing system. All conferees dial the same number and are connected together.

Member—A participant in a system.

Memory—(1) The main high-speed storage area in a computer where instructions for a program being run are temporarily kept. (2) A device in which data can be stored and from which it can later be retrieved.

Menu—A list of choices available to a user that is presented on a monitor. The user selects an action to be performed by typing a letter or by positioning the cursor.

Menu-Driven—A computer system that primarily uses menus rather than a command language for its directions.

Message Switching—Routing data toward its destination. This is done by the computer processor.

Microcomputer—Sometimes called a personal computer (PC) or small business computer. Micros usually support one user, but with increased power may provide processing for several terminals. Physically very small, PCs fit on or under a desk. Microcomputer technology is based on larger-scale integration (LSI) circuitry. Micros are usually the least expensive of the computer types.

Micrographics—Photographic processes by which information can be reduced to a microform medium and be stored and retrieved for reference.

Microprocessor—A single-chip central processing unit incorporating LSI technology. It performs the basic data processing functions of a computer.

Microwave Transmission—Electromagnetic transmission of data, audio, or video signals through open space on a line-of-sight path.

Migration Path—A series of alternatives outlined by a computer manufacturer that enable the user to introduce new computer equipment into a system. It allows an individual to increase a system's computing power by adding or trading in components rather than giving up current hardware and software.

Minicomputer—A type of computer that is usually smaller in size and capability than a mainframe. Its performance generally exceeds that of a microcomputer. Since minicomputers are more modular than mainframes, they can be configured to provide better price/performance systems.

Mnemonics—(1) Groups of letters and numbers that bring material (files or fields of information) onto a display screen. (2) Short, easy-to-remember names or abbreviations. Many commands in programming languages are mnemonics.

Modems (MOdulator/DEModulator)—A hardware device that permits computers and terminals to communicate with each other using analog circuits such as telephone lines. The modem's modulator translates the digital computer signals into analog signals that can be transmitted over a telephone line. The modem's demodulator converts analog signals into digital signals for the computer's use.

Monitor (Hardware)—A television-like display screen that can be used as an output device. It is also called display screen, cathode ray tube (CRT), and video display terminal (VDT).

Monitor (Software)—Part of an operating system that allows the user to enter programs and data into the memory to run programs.

Monitor File—A file that will record all user activities for recovery and accounting purposes, as well as for recordable system errors.

MOS—Metal-oxide semiconductor, the most common form of LSI technology.

Multicopy Form—A preprinted, multiple form that contains carbon paper between the pages (e.g., W2 forms and credit card receipts).

Multidrop Lines—A communication network in which more than one terminal is located on a single line connected to the computer.

Multiplexer—A device that combines streams of information into a composite signal and sends them along a communicating channel. A similar device reverses the process at the receiving end of a transmission.

Multiprocessing—Processing by two or more computers connected to run jobs concurrently for faster results.

Multiprogramming—A scheduling technique that allows more than one job to be executed at any one time. Thus, one CPU can appear to be running more than one program because it gives small slices of time for executing each program.

Multitasking—The execution of several tasks at the same time. Although computers can perform only one task at a time, the speed at which a computer operates makes it appear as though several tasks are being performed simultaneously.

Nanosecond—One billionth of a second.

Narrowband Channel—Usually refers to a telephone circuit that handles 3,000 herz.

Network, Electronic—A group of computers or other devices connected by cables or through telephone lines. The computers send and receive data among themselves and share certain devices such as hard disks and printers.

Node—(1) Department of an organization. (2) A connecting point on a communicating network or between communicating channels.

Nonsimultaneous Communication—Messages being received and transmitted at a terminal regardless of whether a receiver is present.

Nonvolatile Memory—Memory that is not lost when a processor's power supply is shut off or disrupted.

Other Common Carrier (OCC)—Term originally used to describe telephone service companies other than AT&T. OCC now includes AT&T.

OCR—See Optical Character Recognition System. Also called optical character reader.

OEM (Original Equipment Manufacturer)—A middleman who buys computer equipment from hardware and software manufacturers or vendors, and resells or repackages it.

Off-Line—Equipment that is disconnected from a processor. Some can still be operated off-line (independently).

OJT—On the Job Training.

On-Line—The operation of peripheral equipment or devices in a system that is under the CPU's control.

Operating System—Computer programs that allow a computer to supervise its own operations. They accomplish such functions as input/output control, memory allocation, and program read-in e.g., UNIX. Small systems are called monitors, supervisors, or executive programs.

Optical Character Recognition System—A light-sensitive optical scanning system that senses and encodes alphanumeric characters into digital format.

Optical Disk—Computer storage (memory) disk. They have potential for far greater capacity than magnetic disks. Some can be rerecorded.

Option Module—An add-on printed-circuit module that allows expansion of a system. See Board.

Organizational Directory—A directory of a computer conferencing system that contains information relating to its members, centers, and groups.

Outbox—A file or directory containing references to information distributed by a member.

Output—Information produced as a result of processing input data.

Packet Switching—A relatively new form of digital communication in which data bits are grouped into bursts (or packets) of fixed length so they can share a channel with other such bursts. When received at the destination, the bursts are separated and sent to the appropriate recipients.

Parallel Communication—Data transmission in which a number of bits are transmitted simultaneously over separate wires.

Parallel Interfaces—A feature allowing two lanes of data to flow simultaneously along a channel.

Parallel Transmission—Sending more than one bit at a time.

Parameter—A range of characteristics of a program/record or other area.

Parity—A "one-extra-bit" code used to detect recording or transmission errors.

Parity Bit—An extra bit added to a character's binary code to make it conform to the parity checking method (see Parity Check).

Parity Check—A method of error detection in data communications that checks whether the sum of 1 bits in each character received is even or odd. In odd parity, the sum of 1 bits in a character must be odd; if the character's pattern would otherwise be an even number of bits, it is transmitted with the added parity bit set to 1. In even parity, the opposite occurs; the parity bit is set to 1 for characters with odd bit patterns.

Password—A word each user may attach to his/her mailname. Passwords may also be attached to groups and centers for security purposes.

PBX (Private Branch Exchange)—An organization's internal switchboard.

PCM—Depending on the reference, either Plug-Compatible Manufacturers or Pulse Code Modulation.

Peripheral—A device that is external, but connected, to the CPU and main memory of a system. A printer, modem, or terminal would be a peripheral device.

Personal Computer (PC)—See Microcomputer.

Pixels (Picture Elements)—A dot or cluster of dots that form the smallest unit of a picture that is seen on a computer display screen. For graphics displays, screens with more pixels generally provide higher resolution.

Plotter—A graphic drawing device.

Point-to-Point—Place-to-place or station-to-station.

Polling—A method used in data communications networks in which each terminal is asked if there is data to be sent.

Port—An input and/or output socket on a computer that is used to connect hardware such as modems or cables.

Power Supply—It energizes components such as integrated circuits, monitors, and keyboards, and steps down the power supplied to some components.

Printer—Equipment that produces a paper copy of a document (hard-copy output). There are impact and nonimpact printers.

Printhead—The element in a printer that forms a printed character.

Printout—Computer-generated hard copy.

Printout Queuing—A computer function that allows a number of documents to be lined up for printout. Some systems allow the operator to designate the order; others operate on a first-in, first-out basis.

Processor—The controlling unit or processing part of the computer system that reads, interprets, and executes instructions.

Program—The complete sequence of instructions and routines needed to solve a problem or to execute directions in a computer.

Program Disk—A disk containing the instructions for a program.

Programming Language—The words, mnemonics, and/or symbols, along with the specific rules allowed in constructing computer programs. Some examples are BASIC, FORTRAN, and COBOL.

PROM (Programmable Read-Only Memory)—A permanent memory chip for program storage.

Protocol—A set of rules and conventions governing the formats used in data communications.

Protocol Converter—Device for translating codes or protocols between networks or devices.

Public Data Network (PDN)—A packet- or circuit-switched network available to many clients. A PDN may offer value-added services at a reduced cost because of communications resource sharing. It also usually provides greater reliability due to built-in redundancy.

Query Capability—Commands provided for the user to select and retrieve information.

RAM (Random Access Memory)—Memory that can be read and written into (i.e., altered) during normal operation. RAM is the type of memory used in most computers to store the instructions of programs that are being run.

Raster—A computer graphics coding system. The coding represents the dots that compose a picture.

Real Time—The actual time an event is occurring. A term used to describe an on-line interactive application. Some computer conferences can be held in real time.

Recipient—A person who receives mail submitted by an electronic mail system or within a computer conferencing system.

Record—A collection of related data items.

Remote Job Entry (RJE)—Entering jobs in a batch processing system at a location distant from the central computer site.

Remote Terminal—Input/output equipment attached to a system through a transmission network.

Reprographics—Mass reproduction of documents, graphics, and film by such processes as offset printing, photocopying, and microfilming.

Resolution—The degree of detail that can be seen on a display screen.

Resource Directory—An electronic file containing information associated with all private and public permanent storage areas within a computer conference.

Reverse Video—The ability to reverse a standard display on a terminal monitor to highlight characters, words, or lines.

Ring—A circular local area network where messages pass from station to station by passing an access token or by means of a polling technique.

RJ11—A standard modular telephone jack into which a direct-connect modem can be plugged.

RJE—See Remote Job Entry.

Role—A participant's function within a computer conference.

ROM (Read-Only Memory)—Permanent memory written during manufacture, or a permanent memory chip for program storage.

RS-232C—A standard connection for serial computer communications as described by the Electronics Industry Association (EIA). The standard specifies the physical connections between computers and other devices, such as modems and printers, and defines characteristics, such as baud rate, of the electrical signals sent through the connection.

Screen Editor—A feature that supports the cursor concept. The cursor may be moved around the display screen to identify portions of text to be manipulated. Line and character modifications are initiated by special character combinations and blocks are manipulated by entering commands.

Screen Format—An arrangement of characters, numbers, or lines and columns that fill a screen.

Screen Size and Type—The dimensions and kind of screen display (CRT, gas plasma, or LED). A display screen facilitates the job of entering and editing text.

SDLC (Synchronous Data Link Control)—A means of transmitting protocol similar to BISYNC. Data is gathered into blocks. SDLC uses data bits to signal control functions instead of the full characters used by BISYNC. It is part of IBM network architecture.

Search Capability—The methods by which a system searches for an editing point. A system can search by document, page, paragraph, word, or character string.

Sender—Person submitting an electronic message.

Serial Communication—Data transmission in which each bit is sent separately and sequentially.

Serial Interface—A single-channel connection between computers and peripheral drives, printers, and modems.

Serial Printers—See Daisy Wheel Printers.

Shared User—A computer system which shares computer resources.

Soft Copy—Information presented on a display screen or in audio format rather than as printed copy.

Soft Disk—See Diskette.

Soft Keys—Keys on a computer keyboard that can be given special functions or programmed to suit the user's needs.

Software—Instructions that make a computer perform a specific task or program.

Software Interface—A program that controls the way a computer program interacts with other programs it uses.

Sort—Rearranging information that has been filed in "fields."

Sort Keys—Keys that indicate a sequence of information order.

Speakerphone—An amplified telephone that allows hands-free usage.

Standard Member—A standard member in a computer conference participates in the system without a leadership role.

Star Network—A network containing a central computer at the hub. All equipment radiates from that center.

Storage Media Capacity—Maximum number of characters the storage media (such as mag cards, disks, or tapes) can hold.

Storage Media Standard—The design of disks and tapes used to store memory.

String—Alphanumeric data that is treated as a unit.

Subject—A subject is a one-line header used as a summary of the information content of a message, note, or document.

Subroutine—A group of instructions which are used several times in a program and can be called up as needed.

Switched Line—A type of data communications line used to connect computers over a telephone network.

Synchronous Transmission—A method of high-speed transmission in which the timing of each bit of data is precisely controlled.

System Center—A center established during the initiation phase of a teleconference system that will exist during its lifetime.

System Components—The physical parts of a system, such as a keyboard, CRT display, minicomputer, mag card reader or floppy disk drive, and printing device.

System Security—Some systems provide an electronic key lock to prevent unauthorized access. Other systems have security codes that allow only certain persons to access stored documents.

Tape Drive—The I/O unit housing a magnetic tape reel that reads data recorded on tape and records data on the tape.

Teleconferencing—See Computer Teleconferencing.

Teleprocessing—Processing of data that is received from or sent to remote locations by way of telecommunication lines.

Telex, TWX—Switched telecommunication services.

Term Dictionary—A file that stores a technical vocabulary of frequently used words and phrases that are specific to the individual business. They are retrieved by typing a fewer number of characters than would normally be necessary.

Terminal, Data Communication—A terminal used in a data communications system for transmitting and receiving data.

Text Editor—A program that assists in text preparation and editing.

Tieline—A voice or circuit trunk line between two PBXs.

Time Division Multiplexing—A method by which each node is allotted a small time interval during which it can transmit a message or part of a message. In this way, messages of many channels are interleaved for transmission. They are then demultiplexed into their proper order at the receiving end.

Timesharing Operating System—A system that executes a number of processes at the same time though they are controlled from several different terminals.

Token Passing System—A method by which equipment waiting to transmit a message monitors a system, waits for an empty "token" or frame, and inserts a message and its address. When equipment finds a message addressed to it, it retrieves the message and sets the token on empty.

Transceiver—A device that both transmits and receives analog or digital signals.

Transducer—A device that converts sounds waves to electrical signals.

Transfer Rate—The volume of information per time unit that gets transferred between a random access storage device and main memory, or between any two devices.

Transparent—Any function that is invisible to a user.

Typeface—A specific style of print, including alphabet, numbers, and symbols, reproduced by a "font."

Unlisted Member—An unlisted member of a computer conference experiences all the benefits of participating in the system, but his/her participation is not publicly known.

Uploading—Shifting information from memory banks of one computer to another, generally from a PC to another computer.

User Interface—The pathway or connection between a person and device.

Value-Added Carriers—Communications networks that provide additional services.

VAN—Value-Added Network.

Vertical Scrolling—Vertical movement of characters on a display screen that allows more lines to be shown.

Videotex—A service that uses part or all of a TV screen for information displays called pages or frames. The information could range from weather or news to advertising for various services or specialized information.

Virtual Storage—A method in which portions of a program are placed in auxiliary storage until needed. This gives the illusion of unlimited main storage.

VLSI (Very Large Scale Integration)—Refers to microelectronic chips carrying up to 1 million transistors.

Voice Grade Line—A normal telephone line designed for voice communication.

Voice Mail—A system that provides computer-controlled deposit, storage, and delivery of voice messages.

Voice Switching—A systems feature whereby a speaker's voice activates a transmission.

Volatile Memory—Memory that is lost when processor's power supply is shut off.

WAN—See Wide Area Network.

Wide Area Network (WAN)—External dedicated or nondedicated, switched or nonswitched, communications network.

Winchester Disk—A hard disk for computer memory storage.

Word Processing—Using terminal and related storage devices for data storage, manipulation, and processing needed to prepare letters and reports.

Word Wraparound—Equipment automatically places a word on the next line when it cannot fit on the line being typed.

Work Space—A dedicated main memory storage area maintained by the system for each user during a session.

Work Window—Compartments on a monitor that display work or documents. About six windows can be viewed at one time.

X.25—A protocol standard for interface connections between equipment and public data networks.

X.400 Message Handling Facility—A standard that would enable people using various communicating equipment and networks to exchange voice, text, and graphics messages.

Bibliography

Abrams, Bill. " 'Middle Generation' Growing More Concerned with Selves." *The Wall Street Journal,* January 21, 1982, p. 25.

"Again, Pollution Is a Major Concern; This Time It's the Indoor Variety." *Business Facilities,* Critical Issues/Commentary column, July 1985, p. 31.

Allen, Bob. *Trends in Information Management: IRM Long-Range Report* (a student's analysis of his job activities in a brewery), March 7, 1984.

Antonoff, Michael. "The Push for Telecommuting." *Personal Computing,* July 1985, pp. 82–92.

Association of Electronic Cottagers (pamphlet). Sierra Madre, CA 91024.

Atkinson, William. *Working at Home; Is It for You?* Homewood, Ill.: Dow Jones-Irwin, 1985.

_____. "The Psychology of Working at Home." Telecommuting Technology Conference, lecture, 1985, Boulder, Colorado.

AT&T Communications. "Teleconferencing—Where Does It Fit?" July 1985, p. 7.

"Background on the FUNDI." *Public Works Canada,* Fall 1984, pp. 1–2.

Baetz, Mary L. *Planning for People in the Electronic Office.* Toronto, Canada: Holt, Rinehart and Winston, Ltd., 1985.

Bellew, Patricia A. "Technology" (column). *The Wall Street Journal,* May 10, 1985, p. 27.

Boden, Margaret A. *Artificial Intelligence and Natural Man.* New York: Basic Books, Inc., 1977, p. 333.

Branton, P. "Behaviour, Body Mechanics, and Discomfort." In *Sitting Posture,* ed. E. Grandjean. London: Taylor & Francis, 1969, pp. 202–13.

Bratton, E. C. "Concepts of Energy and Work in Home Management." *Journal Home Economics* 51, 1959, pp. 102–4.

British Standards Code of Practice, Angus, T. C. *The Control of the Indoor Climate.* Oxford: Pergamon Press, 1968.

Broadbent, D. E. "Effect of Noise on an Intellectual Task." *Journal Acoustic Society of America* 30, 1958, pp. 824–27.

Bulkeley, William M. "Better than a Smile: Salespeople Begin to Use Computers on the Job." *The Wall Street Journal,* September 13, 1985, p. 25.

Bulkeley, William M. "In the Field, Lap-Top Units Get Data, Print Proposals." *The Wall Street Journal,* September 13, 1985, p. 25.

Chabrow, Eric R. "Telecommuting: Managing the Remote Workplace." *InformationWEEK,* April 15, 1985, pp. 27–35.

Chin, Kathy. "Home Is Where the Job Is." *InfoWorld,* April 23, 1984, pp. 30–36.

Conrath, David W. "Office Automation: The Organization and Integration." *OAC '85 Conference Digest.* Atlanta, Ga.: afips Press, pp. 310–11.

Committee on the Hygiene of Housing. "Construction and Equipment of the Home." Chicago: American Public Health Association, 1951.

Cutler, Ivan. "Designing the Office around the Computer." *Personal Computing,* November 1984, pp. 66–70.

"CW at NCC" (column). *Computerworld.* July 16, 1984, p. 28.

Dallow, Peter K. *Telecommuting in Fort Collins: A Case Study.* October 15, 1985.

Data Communications. "Newsfront" (column). May 1984, pp. 45–48.

Data Communications. "Viewpoint" (column). May 1984, p. 13.

D'Attilo, Lauren. "On the Job." *Datamation,* February 15, 1985, pp. 156–58.

The Denver Post/News Center 4 April (1985) Survey. Boulder, Colo.: Talmey Associates.

Derven, Ron. "Plan Now for Emerging Markets." *High-Tech Marketing,* September 1984, pp. 34–38.

DeSanctis, Gerardine. "A Telecommuting Primer." *Datamation.* October 1983, pp. 214–20.

Downing-Faircloth, Margo. "Would Working at Home Be Wise?" *Personal Computing,* May 1982, p. 42.

Drucker, Peter F. "A Prescription for Entrepreneurial Management." *Industry Week,* April 29, 1985, pp. 33–40.

————. "Playing in the Information-Based 'Orchestra.' " *The Wall Street Journal,* June 4, 1985, p. 28.

Duffy, Francis, et al. *The Orbit Study,* DEGW, England, 1984.

Dutton, William; Janet Fulk; and Charles Steinfield. "Utilization of Video Conferencing." *Telecommunications Policy,* September 1982, pp. 164–78.

Dworetzky, Tom. "Machines on the Go." *Discover,* July 1984, p. 77.

Eisen, Margaret. "Business Computing in the Home—How You Can Make It Happen." *Computer Dealer,* March 1984, pp. 62–68.

"Employee Preference for Working" (chart). Sources include: *The Wall Street Journal,* Honeywell Information Co., Cross Information Co., Bureau of Labor Statistics.

Farkas, David. "White Collars with Union Labels." *Modern Office Technology,* May 1985, pp. 118–22.

"Functionally Diagnosing the Office." *Public Works Canada,* July 10, 1985, pp. 1–4.

"Fundi Project Summary." *Public Works Canada,* Summer 1985, pp. 1–2.

Galbraith, Jay. *Organization Design.* Amosov Park, Calif.: Addison-Wesley Publishing Company, 1977.

Garland, Anne Witte. "The Surprising Boom of Women Entrepreneurs." *Ms.* magazine, July 1985, pp. 95, 106.

"Getting to Work" (Census Bureau chart). *The Wall Street Journal,* March 11, 1985, p. 25.

Glossbrenner, Alfred. *The Complete Handbook of Personal Computer Communications.* New York: St. Martin's Press, 1985, pp. 461–77.

Gluckin, Neil. "The Office Is Where the Workers Are." *Telecommunication Products + Technology,* June 1985, pp. 56–60.

Gold, Elliot. "Attitudes to Intercity Travel Substitution." *Telecommunications Policy,* June 1979, pp. 88–104.

Goleman, Daniel. "The Human-Computer Connection." *Psychology Today,* March 1984, pp. 20–21.

Gordon, Gil E. "Microcomputers Spur Interest in Telecommuting." *Computerworld,* April 29, 1985.

Gordon, Gil. "Telecommuting." *Profiles,* May 1985, pp. 30–76.

Gouin, Michelle, and Thomas B. Cross. *Intelligent Buildings: Strategies for Technology and Architecture.* Homewood, Ill.: Dow Jones-Irwin, 1985.

Goyal, P., and B. C. Desai. "A Personal and Remote-Work Station." *OAC '85 Conference Digest,* 1985, Atlanta, Ga.: afips Press, pp. 25–31.

Grandjean, Etienne. *Ergonomics of the Home.* London: Taylor & Frances, Ltd.; New York: John Wiley & Sons, 1973.

Haas, Charlie. "Elements of Style." *Access* (*Newsweek* magazine), Fall 1984, pp. 76–81.

Harkness, Richard Chandler. *Telecommunication Substitutes for Travel.* Ann Arbor, Mich.: Xerox University Microfilms, 1973.

_____. *Technology Assessment of Telecommunications/Transportation Interactions,* vols. I and II: Stanford Research Institute. Menlo Park, Calif. 1977.

Harper, F. C.; W. J. Warlow; and F. L. Clarke. *The Forces Applied to the Foot in Walking.* National Building Studies, Research Paper No. 32. London: H. M. Stationery Office, 1961.

Haynes, K. J., and J. Raven. *The Living Pattern of Some Old People.* Garston, Watford, Herts, Building Research Station, Miscellaneous Paper No. 4, 1962.

Hellman, Hal. "Home Sweet Office." *High Technology,* February 1985, pp. 64–66.

Hirsch, Phil. "Big Companies Try Voice-Data Nets." *InformationWEEK,* April 15, 1985, p. 64.

Hole, W. V., and J. J. Attenburrow. *Houses and People.* London: H. M. Stationery Office, 1966.

Honan, Patrick. "Telecommuting: Will It Work for You?" *Computer Decisions,* June 15, 1984, pp. 88–92.

Intelligent Buildings and Information Systems—IBIS Executive Report. Boulder, Colo.: Cross Information Company, May 1985.

Intelligent Buildings and Information Systems—IBIS Executive Report II. Boulder, Colo.: Cross Information Company, November 1985, p. 253.

"Interim Design Guidelines for Automated Offices." Center for Building Technology, NBSIR 84-2908, August 1984, pp. 53–54.

Jacobs, Sanford L. "Software Is a Cheap Business to Get into—But Many Fail." *The Wall Street Journal,* April 29, 1985, p. 23.

Jarzab, James T. " 'Home' Computing Perils." *InfoWorld,* Viewpoint (column), February 25, 1985, p. 8.

Johansen, Robert, "Teleconferencing Success Stories." *International Teleconferencing Association (ITCA),* March 1985.

Johansen, Robert; Jacques Vallee; and Kathleen Springer. *Electronic Meetings: Technical Alternatives and Social Choices.* Menlo Park, Calif.: Addison-Wesley Publishing Company, 1979.

Jones, Verna Noel. "Work Industry Sliding into Home Base." *Rocky Mountain News,* Denver, April 6, 1984, p. 51-W.

Kapple, W. H. *Kitchen Planning Standards.* Urbana, Ill.: University of Illinois Small Homes Council C 5.32, 1965.

Kleeman, Walter B., Jr. "The Electronic Office-Design Opportunity." *Electronic Office Design,* p. 8.

Kleeman, Ph.D., Walter; Francis Duffy, Ph.D.; Michele K. Williams, I.B.D.; Kirk P. Williams, I.B.D. *Designing the Electronic Office: A Practical Guide.* To be published in 1986 by Van Nostrand Publishing Co.

Koch, K. W.; B. H. Jennings; and C. H. Humphreys. "Is Humidity Important in the Temperature Comfort Range?" *ASHRAE Transactions* 66, 1960, pp. 63–68.

Koenig, Peggy. "Telecommuting: One Firm's Approach." *CommunicationsWeek,* sec. C, April 8, 1985, pp. 1–2.

Kollen, J. H., and J. Garwood. *Travel/Telecommunications Trade-off: The Potential for Substitution Among Business Travelers.* Montreal: Bell Canada, 1975.

Kramer, Kenneth L. "Telecommunications—Transportation Substitution and Energy Productivity." *Telecommunications Policy,* March 1982, pp. 39–59.

Kuzela, Lad. "More High-Tech Firms Turning to 'Temp' Employees." *Industry Week,* May 27, 1985, p. 33.

Land Use Digest, July 15, 1985. p. 3.

Leinfuss, Emily. "Labor Group Hits VDT Hazards." *Management Information Systems Week,* May 29, 1985, p. 34.

Lewis, Mike. "If You Worked Here, You'd Be Home Now." *Nation's Business,* April 1984, pp. 50–52.

Lindsay, Susan. "Computer Condominium Offered." *Systems Users,* May 1985, pp. 34–35.

Locke, E. A. *Toward a Theory of Task Motivation and Incentives.* Organizational Behavior and Human Performance, 1968, pp. 157–89.

Longhini, Gregory. "Coping with High-Tech Headaches." *Planning,* March 1984, pp. 24–32.

Lublin, Joann S. "Running a Firm from Home Gives Women More Flexibility." *The Wall Street Journal,* December 31, 1984, p. 11.

————. "Small Business" (column). *The Wall Street Journal.* May 28, 1985, sec. 2-p.

Machalba, Daniel. "Like Other Cities, Hartford Has Gridlock; Unlike Others, It's Not Building Roads." *The Wall Street Journal,* April 29, 1985, p. 23.

Madame Mogul (astrology column). *The Courant* (Boulder County, Colo.), June 12, 1985, p. 11.

Maney, Kevin. "Companies Trim Staffs, Fatten Profit." *USA Today,* July 3, 1985, sec. B, pp. 1–2.

Mason, Roy. *Xanadu,* Washington D.C.: Acropolis Books, Ltd., 1983.

Masuda, Yoneji. *The Information Society.* Tokyo: Institute for the Information Society.

McAdams, Susan. "In Search of the 'Urban Telecommunications Infrastructure.'" *Scag Telecommunity,* May 1985, p. 2.

McCullough, H. E.; K. Philson; R. H. Smith; A. L. Wood; and A. Woolrich. *Space Standards for Household Activities.* Illinois Agricultural Experiment Station bull. 686, Urbana, Ill., 1962.

McFarland, R. A. *The Study and Control of Home Accidents.* Boston: Harvard University, School of Public Health.

Micossi, Anita. "The Ten-Second Commute." *PC World,* December 1984, pp. 119–27.

Miller, Arthur. "Are You a Lousy Listener?" *Industry Week,* August 5, 1985, pp. 44–45.

Miller, Michael W. "Technically Speaking." *The Wall Street Journal,* May 20, 1985, pp. 86C–87C.

————. "Computers Keep Eye on Workers and See If They Perform Well." *The Wall Street Journal,* June 3, 1985, pp. 1, 15.

Mintzberg, Henry. *The Nature of Managerial Work.* New York: Harper and Row, 1973.

Morgan, C. T.; J. S. Cook; A. Chapanis; and M. W. Lund, eds. *Human Engineering Guide to Equipment Design.* New York: McGraw-Hill, 1963.

Moser, Penny Ward. "Hooking Up in the Heartland." *Discover,* July 1984, pp. 81–84.

Mullen, Frank. " 'Telecommuters' Break Barriers between Home, Work." *Rocky Mountain Business Journal,* September 23, 1985, pp. 12, 22.

Naisbitt, John. *Megatrends; Ten New Directions Transforming Our Lives.* New York: Warner Books, 1982.

Naisbitt, John, and Patricia Aburdene. "Good News: Here Come the New Mega-trends!" *Ms.* magazine, July 1985, pp. 36–37, 104.

Naisbitt, John, and the Naisbitt Group. *The Year Ahead.* New York: AMACOM, 1985, p. 28.

National Swedish Institute for Building Research. *Study of Dimensions Reequip-ment in Housing for Old Persons.* Report No. 19, Stockholm, 1965.

National Swedish Institute for Building Research, ed. In agreement with the United Nations. *Quality of Dwelling and Housing Areas.* Report 27, Stock-holm, 1967.

Neodata. Interview with assistant to the president. Boulder, Colorado, Novem-ber 1985.

Nevins, R. G.; F. H. Rohles; W. Springer; and A. M. Feyerherm. "A Temperature-Humidity Chart of Thermal Comfort of Seated Persons." *ASHRAE J.* 8, 1966, pp. 55–61.

New York Times. "Flexible Hours." July 2, 1985, p. 20.

Nilles, Jack M.; F. Roy Carlson; Paul Gray; and Gerhard J. Hanneman. *The Telecommunications-Transportation Trade-Off.* New York: John Wiley and Sons, 1976.

Noble, Barbara Presley. "The Venture Survey." *Venture,* January 1985, p. 24.

O'Boyle, Thomas F. "Loyalty Ebbs at Many Companies as Employes Grow Disil-lusioned." *The Wall Street Journal,* July 11, 1985, p. 29.

Ochsman, B., and Alphonse Chapanis. "The Effects of 10 Communication Modes on the Behavior of Teams during Cooperative Problem Solving." *In-ternational Journal of Man-Machine Studies,* September 1974, pp. 579–619.

Olgren, Christine H., and Lorne A. Parker. *Teleconferencing Technology and Ap-plications.* Artech House, 1983.

Olson, Margrethe H. "Do You Telecommute?" *Datamation,* October 15, 1985, pp. 129–32.

Oshima, M. *Optimum Conditions of Chair.* Paper presented at the Fourth Inter-national Congress on Ergonomics, Strasbourg, July 1970, unpublished.

Ostendorf, Virgina A., Interview. Littleton, Colorado, October 14, 1985.

Peelle, Dr. Evan. "How to Make Telecommuting Work." *Personal Computing,* May 1982, pp. 38–40.

Peters, Tom, and Nancy Austin. "A Passion for Excellence." *Fortune,* May 13, 1985, pp. 20–32.

Petherbridge, P., and A. G. Loudon. *Principles of Sun Control. Architect's J.* 143, 1966, pp. 143–49.

Press Information. "Smarthome." CyberLINX Computer Products, Inc. Boulder, Colorado.

Pye, Roger; B. Cartwright; and Hugh Collins. "Prospects for Teleconferencing." *Intelligence Bulletin.* Cambridge: Long-Range Studies Division, Post Office Telecommunications, 1977.

Pye, Roger; Michael Tyler; and B. Cartwright. "Telecommunicate or Travel." *New Scientist* vol. 63, no. 514, 1974, pp. 641–4.

Rothman, David H. "The Computer Cottage Industry Hysteria." *Washington Post*, July 7, 1985, sec. B, pp. 1–2.

Rubin, Arthur. "The Automated Office: An Environment for Productive Work, or an Information Factory." *Executive Summary.* Center for Building Technology/National Engineering Laboratory/National Bureau of Standards, November 1983, p. 10.

Schneider, M. Franz. "The Relationship between Ergonomics and Office Productivity." *OAC '85 Conference Digest.* Atlanta, Ga.: afips Press, p. 126.

Scholberth, H. "Die Wirbelsaule von Schulkindern." *Sitting Posture*, ed. by E. Grandjean. London: Taylor & Francis, 1969.

Schwartz, Lloyd. "U.S. OKs 'Flexitime.'" *Management Information Systems Week*, May 29, 1985, p. 34.

Shirley, Steve. "Why Cottage Industries Fit the Information Age." *Management Technology*, February 1985, pp. 77–78.

Short, John; Ederyn Williams; and Bruce Christie. *The Social Pyschology of Telecommunications.* New York: Wiley and Sons, 1976.

Sigler, P. A. *Relative Slipperiness of Floor and Deck Surfaces.* Washington Bureau of Standards, Report BMS 100, 1943.

Silberstein, J., and F. W. Benton. *Bringing High Tech Home.* New York: John Wiley & Sons, 1985.

Simnacher, Betsy. "Bulletin Boards for Better Business." *Link-Up*, June 1984, pp. 32–34.

Slatta, Richard, Ph.D. "The Problems and Challenges of the Computer-Commuter." *Link-Up*, June 1984, pp. 36–39.

Sloan Management Review vol. 26, no. 2, pp. 45–49.

Steidl, R. E., and E. C. Bratton. *Work in the Home.* New York/London/Sydney: John Wiley & Sons, 1968.

Steinfield, Charles. *The Nature of Electronic Mail Usage in Organizations: Purposes and Dimensions of Use.* Abstract presented to the International Communication Association, San Francisco, May 1984.

Stone, Philip J., and Robert Luchetti. "Your Office Is Where You Are." *Harvard Business Review*, March–April 1985, pp. 102–17.

Svenning, Lynne, and John Rushinskas. *Predicting Receptivity and Resistance to Video Conferencing.* Paper presented at the International Communication Association, Minneapolis, Minn., 1981.

Swartz, Herb. "Telecommuting's Expansion Poses Host of Legal Questions." *InformationWEEK*, April 15, 1985, p. 38.

Tapscott, Henderson, and Greenberg. *Planning for Integrated Office Systems*, Canada: Holt, Rinehart, and Winston, Ltd., 1985.

TC Report (newsletter). Vol. 1, issue 6, June 1985, Electronic Services Unlimited. New York.

Telecommuting Review: the Gordon Report. vol. 1, no. 1, Gil Gordon Associates, Monmouth Junction, N.J., October 31, 1984, pp. 1–10.

Telecommuting Review: the Gordon Report. vol. 2, no. 2, March 1, 1985, pp. 1–18.

Thiberg, S. *Anatomy for Planners.* Parts I–IV. Stockholm: Statens Institut for Byggnadsforskning, 1965–70.

Tyler, Michael; Martin Elton; and A. Cook. *The Contribution of Telecommunications to the Conservation of Energy Resources.* OT Special Publication 77-17, Office of Telecommunication, U.S. Department of Commerce, Washington, D.C., 1977.

Video Display, Work, and Viscon. Washington, D.C.: National Academy Press, 1980.

Wakin, Edward. "Jobs a la Carte." *Today's Office,* September 1984, pp. 43–47.

Wald, Matthew L. "The Smart Building." *The New York Times,* May 12, 1985, sec. 12, p. 1.

Walker, K. E. "Homemaking Still Takes Time." *Journal of Home Economics* 61, 1969, pp. 621–24.

The Wall Street Journal. February 5, 1985, p. 1.

The Wall Street Journal. "Working Women" (chart). March 8, 1985, sec. 2, p. 1.

The Wall Street Journal. "Who Told You That?" (chart). May 23, 1985, p. 37.

The Wall Street Journal, "Home or Office?" (chart). June 3, 1985, p. 23.

The Wall Street Journal. "Tell Me More." (Source: The Hay Group for Management Database), August 9, 1985, p. 19.

The Wall Street Journal. "Smoking Policies" (graphic). September 3, 1985, p. 23.

The Wall Street Journal. "Low Morale." Equitable Life Assurance Society chart, September 9, 1985, p. 23.

The Wall Street Journal, "Want to Move?" (chart). October 8, 1985, p. 33.

Walter, Stephanie K. "How to Avoid the 'Coffee Clash.' " *Management Technology,* March 1984, p. 72.

Wang Laboratories, Inc. *On Human Factors.* Lowell, Mass., 1983.

Ward, J. S., and N. S. Kirk. "Anthropometry of Elderly Persons." *Ergonomics* 10, 1967, pp. 17–23.

Watkins, Dan. "Employee leasing" (letter to the editor). *Business Journal,* May 27, 1985, p. 2.

Webb, Marilyn. "Life in the Electronic Cottage." *Working Woman,* December 1983, pp. 106–9.

Weidlein, James., pres., Information Design. Interview. Boulder, Colorado, June 5, 1985.

Weinberg, Gerald M. "The Squabble over Sunlighting." *Management Technology,* March 1984, pp. 71–72.

Weston, J. R., and C. Kristen. *Teleconferencing: A Comparison of Attitudes, Uncertainty, and Interpersonal Atmospheres in Mediated and Face-to-Face Group Interaction.* Department of Communications, Canada, 1973.

Westrum, Ronald. "Communication Systems and Social Change." Ph.D. dissertation. University of Chicago, 1972.

Wiegner, Kathleen K., and Ellen Paris. "A Job with a View." *Forbes,* September 12, 1983, pp. 143–50.

Willoughby, Kenneth. Letter. Address: Box 317, Fairacres, NM 88033, Compuserv ID# 71565,2005 for EMAIL.

Woldenberg, Jeanne. "Telecommuting: No Workplace Like Home." *Words,* June-July 1984, pp. 24–27.

Wolfgram, Tammara H. "The Right to Choose Where You Work." *Profiles,* May 1985, pp. 38, 60.

Yamaguchi, Y., and F. Umezawa. "Development of a Chair to Minimize Disc Distortion in the Sitting Posture." Paper presented at the Fourth International Congress on Ergonomics, Strasbourg, July 1970, unpublished.

The Yankee Group. *Yankee Ingenuity*™ newsletter, vol. 7, no. 3.

Index